DAUGHTERS OF THE MOUNTAIN

DAUGHTERS OF THE MOUNTAIN

WOMEN COAL MINERS IN CENTRAL APPALACHIA

SUZANNE E. TALLICHET

THE PENNSYLVANIA STATE UNIVERSITY PRESS
UNIVERSITY PARK, PENNSYLVANIA

Publication of this book has been aided
through funding by the
West Virginia Humanities Council.

Library of Congress Cataloging-in-Publication Data

Tallichet, Suzanne E.
Daughters of the mountain : women coal miners in central Appalachia /
by Suzanne E. Tallichet.
p. cm.
(Rural studies series of the Rural Sociological Society)
Includes bibliographical references and index.
ISBN 0-271-02903-X (cloth : alk. paper)
ISBN 0-271-02904-8 (pbk. : alk. paper)
1. Women coal miners—West Virginia.
2. Sex discrimination against women—West Virginia.
I. Title.
II. Series.

HD6073.M62U68 2006
331.4'822334097544—dc22
2006007763

Printed in the United States of America
Published by The Pennsylvania State University Press,
University Park, PA 16802-1003

The Pennsylvania State University Press
is a member of the
Association of American University Presses.

It is the policy of
The Pennsylvania State University Press
to use acid-free paper. This book is printed on
Natures Natural, containing 50% post-consumer waste, and
meets the minimum requirements of American National
Standard for Information Sciences—
Permanence of Paper for Printed Library Material,
ANSI Z39.48–1992.

CONTENTS

PREFACE

Sometimes it is life's tragedies and the sorrow they bring that ultimately provide the personal spark for scholarly investigations, such as the one represented in this book. Ironically, where one woman's life ended, an important part of mine began. The story behind this study of women miners in southern West Virginia started in State College, Pennsylvania, on the morning of October 6, 1979, when I picked up a folded edition of the local paper from the foyer floor. The front-page headline of the *Centre Daily Times* read: "Miner Dives to Safety, But . . . Tragic Scene Etched in His Mind." Being a graduate student in journalism back then, I began reading this article to see how it was written. But the more I read, the more I was drawn into the story itself. A few days earlier there had been an accident at the Rushton Mine in Osceola Mills, Pennsylvania, only about an hour's drive west from our campus community. Far removed from the relatively insular environment of a university town, news of this tragic accident had come spilling out of a dark and dangerous world much like the thirty-foot layer of rock that had descended upon a miner running a roof bolter and his helper, a woman miner named Marilyn McCusker. He escaped certain death; she did not. McCusker had become wedged underneath the slab and had suffocated to death. Horrific as this was, I was just as struck by her co-worker's explanation of events. He said that women didn't belong in the mines because they did not react as quickly as men. Nonetheless, he concluded with the haunting and contradictory comment that "she seemed to hesitate when that rock started coming. Maybe she waited to see if I made it."

As much as McCusker's death had nagged me since that crisp fall morning, it had also fueled my fascination with women in mining. During the years that followed, I read numerous magazine and newspaper articles about women miners, laying the foundation for my doctoral dissertation in rural sociology at Penn State. I knew that women miners were having trouble advancing to more skilled jobs, but I needed to find empirical support for this. So I began with a Bureau of Mines data set based on a random survey of men and women miners nationwide. It included all the variables necessary for establishing job-level sex segregation among underground coal miners and for assessing the effects of gender and human capital factors (age, training, and job experience) on miners' job rank. I found that women were concentrated in the lower job ranks, relative to men, but also that gender was a more powerful predictor of a miner's job rank than all the human capital factors combined. However, confining my inquiry to the use of a variable did not tell me *how* gender was adversely affecting women's advancement. Statistics in no way revealed women miners' lived experiences. I needed to use qualitative techniques, but actually going into the field to interview women miners was a whole other prospect. No one in my family had ever been a coal miner. I did not even know any coal miners. So I began poring over books about mining technology, labor history, and anything else written about women coal miners.

I learned from my reading of the Coal Employment Project (CEP), a women miners' advocacy group. Before going into the field I attended two of their annual conferences and spent hours talking to women miners from across the country, among them Cosby Totten and Bernice Dombrowski. Cosby, a former miner and CEP director, always made me feel included at the women's conferences. And Bernice, who had worked with Marilyn McCusker at the Rushton mine, gave me some pointers about asking questions. Next I needed to locate a cohort of women miners, but where? At the next CEP conference I talked with former coal miner Marat Moore, who was now a journalist and photographer for the *United Mine Workers' Journal*. She suggested several large coal mines for potential study. But how could I get permission to do the interviews, examine employment records, or go on underground tours of the mines? Back on the Penn State campus, mining engineering professor Stanley Suboleski helped me get a coal company's permission to do all these things if I promised to preserve the anonymity of my sources. I have kept this promise and refer to the women I interviewed or conversed with briefly by pseudonym only. I am also deeply indebted to my mentors, Carolyn Sachs and the late Ken Wilkinson, for their sincerity and wisdom during the dissertation phase of the study, and to my colleague Rosalind Harris, who has been a constant source of inspiration and an unflagging supporter of this project.

I recall vividly the gray overcast day I left the cornfields of central Pennsylvania for the coalfields of southern West Virginia. Armed with my interview questions and informed consent forms, a new handheld tape recorder, and several boxes of blank tapes, along with my clothes, some books, and a small wad of traveler's checks, I felt as if I were plunging into another world, excited about who I might encounter yet anxious about what I might not accomplish. For more than a month I was to become a part of a small coal-mining town, yet remain separate from it as I conducted interviews, eyeballed documents, and toured both surface and underground mining facilities.

My first foray into the field was to be nothing less than a transcendental experience. At times it took on an almost surreal quality, because I spent almost as much time observing myself as I did observing others. That was in the fall of 1990. Five years later, after becoming an assistant professor, I returned to the study site during the summers of 1995 and 1996.

My fieldwork began during my between-shift meetings with the women in their bathhouse. I was also a frequent visitor in the lamphouse where miners congregated. When the atmosphere seemed right, I lingered in offices and in local diners to join conversations that sooner or later turned to mining issues. Building rapport with women and men miners was crucial to my study, and at times it came slowly and awkwardly. Some members of the mining community were very open and outspoken. Others were initially suspicious and voiced their fears about keeping their well-paying jobs. They usually declined to be interviewed. For those I did interview, my questions brought back good and bad memories alike. So I learned to tread lightly around miners' unresolved conflicts and resentments. I heard stories about strike violence and miners' mistrust of both the company and their own union. Metaphorically speaking, being in the field was sometimes like being in a gassy mine. A wrong move on my part could spark an explosion, with disastrous results for all involved. Nonetheless, my own privilege became apparent to me: I could always leave town, but those whose lives I had affected could not. Thus, I felt a particular obligation not to stir up trouble or betray anyone's trust. And so, when some women said they feared a backlash if I interviewed their male co-workers and bosses, I chose not to contact them. Although this is the study's most obvious limitation, it is one compromise I made with no regrets and would readily make again.

This book is about the working experiences of a cohort of women miners spanning a twenty-year period at a large underground coal mine in the heart of central Appalachia. In my writing about these women, I became all too aware of the pitfalls that plague researchers who do fieldwork to gather qualitative data. I have tried to avoid essentializing the women I interviewed by recognizing their individuality and the differences between them. I have also tried to avoid romanticizing them. In her work on rural women workers in a western North Carolina mica factory, anthropologist Mary Anglin has aptly stated that she viewed her subjects as "neither militant heroes nor as the pawns of oppressive bosses" but rather as "people laboring with dignity and determination amid difficult circumstances" (2002, 14). Likewise, I saw the women at the mine as agents acting to change their own lives as much as those around them acted to "keep them in their place."

I was surprised to find that these women failed to see themselves as extraordinary, that they underestimated the importance of what they had done, and that they were unsure of their place in history. But if you asked what mining has meant to them, they would tell you that they moved a mountain of social stigma and superstition by descending into its literal depths to toil in the face of constant physical danger, despite unwarranted attempts to erode their dignity. In this sense, I believe these women, these daughters of the mountain, are descendants of a mining legacy, and that their pain and sorrow, joy and triumph, should never be forgotten. In that regard, I hope this book is as much for them as it is about them.

ACKNOWLEDGMENTS

The seventeen-year project upon which this book was based was made possible by the time, talents, and various kinds of support given to me by numerous individuals and institutions.

At Morehead State University, I wish to thank my department for a one semester sabbatical and the Research and Creative Productions Committee for the two grants that supported my postdoctoral research. But what I cherish and remember best have been the kind inquiries about my work on this project made by students in my Appalachian sociology seminars over the years.

I also want to acknowledge the support and inspiration of friend and colleague Chris Hensley and to thank Mary Koscheski for her careful editing and sympathetic reading of an important draft of the book.

Likewise, I will never forget the women miners at Coal Employment Project conferences, namely, Bernice Dombrowski, Marat Moore, and Cosby Totten, whose interest in my research helped fuel my own determination.

At Penn State University, where the project started, my ideas and endeavors were heavily nurtured by mentors and friends Carolyn Sachs and Rosalind Harris.

Numerous other women scholars in the Rural Sociological Society have responded with enthusiasm and encouragement to my work on women miners over the years: Cornelia Flora, JoAnn Jaffee, Sally Maggard, Cynthia Struthers, and Julie Zimmerman. This is also a perfect opportunity to remember the lives of two giants in rural sociology, Kenneth Wilkinson and Janet Fitchen, for their dedication to rural people and rural places.

Thanks go to Rural Studies Series Editors Leif Jensen and Claire Hinrichs and to former Penn State Press editor-in-chief Peter J. Potter, as well as Press staff members who brought this work to fruition. I am equally grateful to the West Virginia Humanities Council for its financial backing of the book.

And because I recognize this research as a special part of my own lived experiences, I have often drawn upon my mother's nontraditional experiences in service to her country as a WAC officer during WWII. Her wisdom and insights lead me to the threshold of this work.

Finally, I will be forever indebted to those women miners who dug coal for a living and who took the time and effort to share their stories with me.

INTRODUCTION

When you ask her why she wanted to be a coal miner, her eyes glisten like the shiny pinpoints of light that dance off a coal seam when caught in the penetrating beam of a miner's cap light. Mining is in her blood now, she says in a low voice, gazing down at her calloused, worn hands, thick from almost two decades of what miners call "brute work." She explains how she barely finished high school, married young, had several children, and, after a bitter divorce from an abusive husband, was left nearly destitute. At first she worked at a hot dog stand at a drive-in theater, a bakery, a restaurant, a grocery store, but none of these jobs paid a living wage or provided health insurance. Even her nursing job, a job that required training, paid only $3.35 an hour, and she had to draw food stamps to survive. She recalls growing up with a single mother, wearing hand-me-downs, being on welfare, and enduring her classmates' taunts about being poor. She vows that her children will never know the humiliation she felt at such a young age.

But growing up in a large family also primed her for a life of hard work. Back in 1975 she heard that a few older married women in a neighboring town had gotten jobs in the mines. These were the best-paying jobs to be had. So she went to night classes to become a mining trainee and, like the men in her class, put in her application at a large local coal mine. They were hired; she was not. Then, a year or so later, things changed. A lawsuit against the coal companies, settled between 1978 and 1979, opened up mining jobs for women throughout the Appalachian coalfields. That fall the company called her in, paid her thousands of dollars in back wages as part of the settlement, and gave her a job. By that time, she was more than ready to go to work at a mine already known as "a woman's mine" because almost a hundred of the eight hundred miners working there were women.

Her grandfather said she'd never make it. But he had been a miner before mechanization, during the hand-loading days when mules brought out the coal. Her brothers were opposed to her new job. Couldn't she find something cleaner? But, as the years passed, their attitudes gradually softened, and sometimes they bragged about their coal-mining sister. Weeks after she went into the mines, her live-in boyfriend left her. He'd had enough of his buddies' teasing about not being able to take care of "his woman." A year later he was back, telling his friends that "she had a mind of her own." Her mother begged her for years to get out of mining, saying it was unsafe for women to be working underground. It's not safe for any of us, man or woman, she told her mother. Besides, we've got to make a living, don't we? When one of her daughters declared that women didn't belong in the mines, she reminded her of the clothes on *her* back growing up. Only her father understood, and he told her to go into the mines while there were still some jobs left. He knew what her struggle was all about and that being financially independent would be a source of pride that she had never really known before.

Like any "red cap" trainee, she knew that she would have to prove herself. Her advice to any new miner is to "take the toughest job underground and go at it." She knew that without a reputation for hard work other miners wouldn't want her on their crew and the union wouldn't back her, regardless of her complaint. She began her career as a general inside laborer or "GI." Getting her "mining papers" and donning a black cap was harder than she anticipated. The work of lifting forty-pound bags of rock dust, oversized blocks, water lines, and heavy-duty power cables, hanging "rag" (ventilation curtain), setting timbers, and shoveling belt lines was rough on her body. But while her physical adaptation was inevitable, she knew that the social adjustment would be even more challenging. The mine and its dangers didn't really scare her. She was much more apprehensive about working with the men. "There was some ornery people there, you know. There really was," she says.

At first, some of the miners were hostile, remarking, "You're a woman and you're taking a man's job." She had retorted, "Listen here, buddy, it was a man that put me here! You know, where I had to work and raise my kids." That was something the men understood, and they never said those things to her again. Most of the time she tried to understand the men's point of view. "I knew it was a man's world when I went in it," she admitted, but she had just as much right to work there as they did. "It's my job and I'm keeping it," she would tell them. Otherwise, the men were usually fun-loving and helpful, but from time to time a few had propositioned her. "They'd tell you how many inches they had and say, 'I bet I can satisfy you.'" So she reminded them of their wives to get them to leave her alone, because aboveground things were different. If you ran into them in the store, she explained, "they would go plumb to the other aisle to avoid you because their wives were with them. The wives were so jealous that the husbands would actually go home and tell them that they didn't work with women." To many miners' wives, the women who worked with their husbands underground in the dark simply had to be "whores." A few of the "bosses chased the women, too," she says. Once, a boss pressured her to have an affair with him. Some women would have affairs to get the easy jobs, "so I had to work like a brute because there wasn't nothing there I would have an affair with." Still, she believed she was fortunate, because other women who had turned down a boss "got the dirty jobs." If that's what it took to get an easier job, she vowed, "I'll take the dirty job every time." Over the years she established herself each and every day as a hard worker, telling the men with whom she worked that she was there "for finance, not romance."

During most of her two decades underground she had been a GI, but off and on she had held several different and sometimes higher-paying jobs. "I was a mason. I run a buggy. Then I went to the plow. Then I cribbed the tail entry. I've run coal. I've run a scoop. I've pinned top. I have done everything that there is to be done except run a miner," she says. If there was "down time," some men were willing to show the women on their crews how to run equipment. One time, one of them did say, "If I teach you my job, you'll end up taking it someday." Her experiences had taught her that some men, the "smart alecks," resented women who could outdo them. This "same bunch" had always told her that there were certain jobs a woman could do, but that running machinery was not for her or any other woman. They'd say, "Send the hens or the girls" to set timbers or to shovel the belts. "It's like they're doing the women a favor, they don't need to be on the equipment." But she always thought that "the equipment is safer than what the men sent the women to do." When some of the bosses made the job assignments, they must have had the same attitudes about women running machinery. She knew that the only way to get ahead was to push, to become a "fighter." Even though over time there were fewer better-paying jobs available, she was willing to bid on a coal-running job. But not all the women were like her. Some were intimidated by the prospect of trying something new and failing. These women preferred to stay on their shift as GIs, where they did not risk making a mistake that "you could never live down."

Every few years the company realigned the miners, assigning them to new jobs as it moved its mining operations from one section to another. When that happened, some men and most of the women were demoted to lower-paying jobs requiring more brute work. But, because management had the contractual right to do so, there was very little the union could do about it. In fact, as the union officials told the miners, most of the grievances they could file had already been litigated and precedents had already been set. As most miners would tell you, the union had operated from a position of weakness for years, and few of them felt completely supported. Out of several hundred of them, only about thirty ever attended the monthly meetings. Those who refused to go anymore were disillusioned and turned off by all the "fussing and fighting over stupid old things. It wouldn't be over no grievances," she says, and besides, company and union officials seemed to be pretty "buddy-buddy." Men and women alike felt that union officials had been bought off by management. Even so, her loyalty, like that of most miners, would always be with the union. None of us should ever forget, she says, shaking her head, that "being union means you have more rights than if you're nonunion."

Today, she is proud of what she has accomplished and reflective about how mining has changed her. "Mining was a whole new world," she declares. "They've worked everybody [both women and men] pretty hard, and the work is still hard. We've adapted." She believes that when women first went into the mine, they were scrutinized more than the new men because their capability was, and has remained, in question. But, beyond proving themselves to the men, she thinks that "each of us has proved to ourselves that we can do our jobs." It has given her the confidence to speak up. "I was never like I am now. When I first came in there I was totally, no joke, nothing, and it was hard. I had some hard feelings, but I loosened up after three or four years. Now, I'm a joker and a prankster," she admits with a pirate smile. "I didn't cuss when I started working in the mines, but [later] I cussed them. When they made it hard on me, I gave it right back." She learned that among miners, "you've got to be able to take it in if you dish it out. You can't be dishing it all the time." Looking back, she says, things have settled down a lot since those first few years. She smiles contentedly when saying that she and her crew are "just like a family, one big family, really. Everybody's working to help each other," she says, adding, "the men I work with got a lot of respect [for me]." Then, shifting her weight in her chair as if to signal her final thought, she says in the same low voice she used earlier, her eyes twinkling once again, "We've grown on each other, we've been in there so long. I guess they kind of got used to us."

WOMEN IN COAL MINING

The foregoing account is an amalgamation of the stories taken from my interviews with fourteen coal-mining women working in southern West Virginia during the early to mid-1990s. To begin to understand why their employment is so momentous, we need only consider a brief history of women in underground mining. In preindustrial times, American women toiled alongside their fathers, brothers, and husbands to extract coal for household consumption. Documents from the early seventeenth century show that female slaves were put to work in a Virginia coal mine (Lewis 1987). With the dawning of industrialization, superstitions about women in the mines bringing disaster and death were well known, and seventeen states passed laws barring women from working underground (Moore 1996b). Challenges to these laws were few and violations were numerous, as women continued to work underground in family-run mines, particularly in the Appalachia region.[1] However, despite vigorous opposition from the miners' union, the United Mine Workers of America (UMWA), women were allowed into the mines during the Depression and World War II thanks to the shortage of

1. With the increased interest in women in mining in the past three decades, numerous stories of women working in the mines in the 1930s and 1940s have come to light. Consider, for example, the story of mining pioneer Ida Mae Stull, who began mining coal with her father near Cadiz, Ohio, when she was six years old. Like other women, she continued to mine despite a state law forbidding it. One of her experiences is worth recounting due to its humorous, if not heuristic, quality. Early in February 1934 Ida Mae and some other women were mining coal when a federal mine inspector came to the mine. While the other women hid from him, Ida Mae kept working. "I knew he was coming to put me out, so I put some rotten eggs in my coal cart. When he came in, I let him have it." Although the inspector was successful in banning her from the mine, the following year her attorney succeeded in overturning her eviction order. Ten years later, she retired from mining at age forty-four (*United Mine Workers' Journal* 1982b).

male labor (President's Commission on Coal 1980). But it was not until July 1973 that the first woman miner crossed a West Virginia coal mine's portal without breaking any federal or state law.

During the mid-1970s coal companies hired only a few women at a time, and usually only when pressured by state enforcement agencies. That changed dramatically in 1978, when a Tennessee-based women's advocacy group, the Coal Employment Project (CEP), won a massive class-action suit filed through the Office of Federal Contract Compliance Programs (OFCCP) against 157 coal companies over sex discrimination in hiring.[2] As part of the settlement, these companies paid back wages

2. Early advocates of women in mining relied heavily on previous legislative efforts to curtail employment discrimination in the industry, specifically Title VII of the 1964 Civil Rights Act and Executive Order 11246, issued one year later (Thrasher 1981). According to Weatherspoon (1985), the act prohibited sex discrimination in employers' recruiting, training, hiring, and promotion practices. It also created the Equal Employment Opportunity Commission (EEOC), which was originally charged with administering Title VII through negotiation and conciliation and with providing fair employment guidelines to employers. However, the act as amended by the Equal Opportunity Act of 1972 also gave the EEOC the power to order appropriate affirmative action, reinstatement, and back pay, and to initiate court action against offending employers.

The purpose of Executive Order 11246, issued by President Lyndon Johnson in 1965, was to give the Civil Rights Act teeth. The new concept of affirmative action (AA) introduced by the executive order was supposed to minimize the more subtle and covert forms of discrimination. It meant that employers were now required to take active steps toward implementing equal employment opportunity and that passive support of such measures was no longer enough. For example, where recruiting new workers was concerned, employers could no longer rely on word of mouth to advertise job openings but were required to announce such openings officially and to remove any artificial barriers to hiring minorities, such as examinations or educational credentials, not necessary to the operation of the business organization.

The antidiscrimination statutes contained in the order applied to employers having $10,000 or more in contracts with the federal government. Moreover, government contractors and subcontractors with fifty employees or more and $50,000 in contracts were also required to file a written affirmative action plan with the EEOC. According to Section 202, the affirmative action plan was to cover employment, training or apprenticeship selection, demotion, upgrading transfer, and pay or other forms of compensation where racial and, later, as the order was amended, sexual discrimination might be occurring. The secretary of labor was charged with administering the order, while the Office of Federal Contract Compliance Programs (OFCCP) was responsible for levying sanctions against employers who failed to comply.

Affirmative action also involved taking measures to compensate women and other minorities for past discriminatory practices by public or private employers. Executive Order 11246 required employers who had discriminated against racial and sexual minorities to set hiring goals for reducing and eventually eliminating such discrimination. A discrepancy in the number of minorities in a given job category compared with population estimates of the number of qualified applicants who could have been employed in the same capacity constituted evidence of discrimination. The offending employer's hiring goals would then be set to

to the women they had failed to hire. Not surprisingly, 830 women were hired to work underground in 1978, compared to 19,036 men (Moore 1996b). By the early 1980s, 3,871 women were working underground, about one-third of them in West Virginia mines (Hall 1990). At that time they represented slightly less than 3 percent of all underground miners nationwide (U.S. DOE 1987). In the following decade the coal industry went into decline, and many of the women with "mining in their blood" were laid off with little hope of ever returning to the mines. Notably, most of the women I interviewed for this study were hired as a result of the suit, and several of them were paid back wages. All were a part of the largest and most sustained cohort of women ever to work in the nation's underground coal mines (Moore 1996b).

It is relatively easy for anyone to appreciate, intuitively, the uniqueness of women working underground. But what makes them worthy of study? After all, their numbers are small and their employment spans only a little more than three decades. It has been suggested that their success in breaking legal barriers and old taboos has been more symbolic than anything else (Tickamyer and Henderson 2003). I contend, however, that the full cultural impact of women in mining remains to be seen because their working lives exemplify the direction and ongoing nature of social change. This study highlights the individual setbacks they suffered and the social price they paid for having the temerity to instigate that change. It also identifies these women's successful accommodations and resistance to male and capitalist domination. Understanding their experiences and their challenge to male privilege is both instructive and inspiring to women currently working in any male-dominated industry. More specifically, this study also contributes to our knowledge of rural women's employment in Appalachia. One need only review the work on rural Appalachian women to see that they have routinely been left out of studies of the region (Beaver 1999; Seitz 1995; Smith 1999). For example, Beaver (1999, xvii) has written that women "receive short shrift in Appalachian history, as silent or invisible bystanders to, or even nonexistent in, the acts and events of white men." Smith (1999, 4–5) notes that "the chief protagonist in the history of Appalachia, at least as it's been written in the past 30 years, is a valiant working-class (or small-land holding) man, who struggles against plant-

close the numerical gap by giving qualified women and minorities preference for filling certain jobs where they previously had been underrepresented.

ers, land speculators, coal operators, condescending missionaries, local colorists, and disparaging academics to assert his dignity and power." Similarly, Seitz (1995, 5) has declared that even though "feminist discourse has opened to other voices, it has not yet listened to Appalachian women."

While some recent studies have examined rural Appalachian women's working lives in the informal economy or in service industries, or both (Maggard 1994, 1999; Oberhauser and Turnage 1999; Seitz 1995), rural Appalachian women's experiences working in nontraditional or blue-collar industries continue to be relatively understudied. Although the social and economic changes associated with the creation of an industrial labor force in Appalachia have received considerable attention from scholars, gender relations have often been neglected (Oberhauser and Turnage 1999, 113), including those occurring between women and men working underground. While the more recent work on rural Appalachian women examines their lives as mothers, sisters, wives, and daughters of coal miners (Couto 1993; Gibson-Graham 1996; Giesen 1995; Maggard 1998; Seitz 1995; Scott 1995), there have been virtually no recent studies of Appalachian women *as* coal miners.[3]

All told, I spent two months in the field interviewing fourteen coal-mining women and conducting both formal and informal interviews with additional women and with male miners and the mine superintendent. Seven of the interviews with the women were repeated between the 1990 and the 1995–96 data collection periods, allowing for some comparisons. I was able to glean additional information about miners' employment histories from company documents posted in the lamphouse at the mine. My almost daily visits to the mine, particularly the women's bathhouse, an eight-hour underground tour of the mine, and my forays into the local community yielded copious field notes about the

3. Despite the publicity about their struggles in the male-dominated world of mining, only a few studies have ever documented Appalachian women miners' working lives. In 1978 Mahoney described the entry-level experiences of fifteen women coal miners working at coal mines in two Appalachian states. Using a snowball sample of twenty-five women miners, Hammond and Mahoney (1983) investigated the advancement problems among women working at different mines in central Appalachia. Similarly, Yount (1986) focused on the barriers to women's advancement underground based on interviews with women and men miners in three western states. In 1981 the Coal Employment Project surveyed women miners in the Appalachian region about their jobs and experiences with sexual harassment (White et al. 1981). Most recently, Moore (1996b) spent ten years gathering oral histories from three generations of women miners in the nation's coalfields.

relationships between women and men miners and their supervisors at the mine, and about members of the local community in general. Virtually all of the miners I interviewed were lifelong residents of southern West Virginia who were born into coal-mining families. Twelve of the women were white; two were black. Most were in their thirties or forties and had at least a high school diploma. Like the jobs they performed underground, their experiences with men and mining also varied. (See the Appendix for a discussion of the techniques used, the job ladder, and a detailed profile of the sample.)

Because these women miners were affected by the interdependent forces of patriarchy and capitalism, I used a socialist feminist interpretation. Their dual subjugation gave them particular ways of viewing their world and their status in it. I expanded this interpretation to include race and the colonization process whereby working-class residents of the Appalachian region were further exploited and subordinated by outside capitalist interests, both economically and culturally. While Appalachian women miners' situated knowledge is predominantly a function of the intersection of gender and class, it is also shaped by the interrelated social constructions of race and region. Thus, the gendered interests of women in the sample were occasionally splintered by racism but were much more often exacerbated by conditions confronting them as working-class residents of colonized rural Appalachia.[4]

4. Regarding the "privileged" nature of women's situated knowledge, Sachs (1996) raises a particularly pertinent point for my discussion in her review of feminist standpoint theory. She points out, as Haraway (1991) has asserted, that although there are many good reasons for trusting the subjugated knowledge of women, one must avoid the pitfalls of romanticizing and uncritically accepting their views. In other words, while these subjugated standpoints are rich, important, and even preferred because of where they come from, they are no less vulnerable to various distortions. Thus their situated knowledge is also partial knowledge (see Haraway 1991). Nonetheless, Sachs (1996, 17) concludes that "Haraway's concept of situated knowledge proves particularly helpful in grappling with questions relating to rural women." In addition, during the past two decades several feminist scholars have moved beyond the standpoint concept. In particular, Haraway has argued that there is no universal women's standpoint. Rather women possess multiple and therefore partial standpoints based on where they are situated socially. Thus the concept of situated knowledge is used in this study within the context of standpoint theory, because not only are rural Appalachian women situated differently from other women, women miners are situated differently from other rural women in the same region, such as miners' wives. Likewise, white and black women are situated differently. Yet, at the same time, these diverse groups of women are all subject to patriarchal control and the subsequent division of labor underground, just as they have been in their homes and communities. On the one hand, as Haraway would argue, the knowledge of these women is subjugated knowledge. On the other, the differences in their situated knowledge may explain the differences in their respective strategies for and modes of resistance to patriarchal control. While it is both prudent and advantageous to acknowledge

Other researchers who have focused on the lives of rural women also recognize "the diversity that exists spatially and culturally, in the intersections with other social locations, and in the choices and constraints available to women in their allocation to various roles" (Tickamyer and Henderson 2003, 109). We need to explore and evaluate these intersections. Collins (1990, 225–27) conceptualizes these intersections as "interlocking systems of oppression" in the "matrix of domination" consisting of gendered, racial, and class-based power and privilege in a multiplicative rather than an additive way. Because individuals are simultaneously members of multiple dominant and subordinate groups, at any given time or place they can be either oppressors or oppressed. There are three overlapping sites or levels of domination and subordination—the personal or biographical; the group, created and experienced culturally in terms of gender, race, and class; and social institutions. These systems of oppression reinforce one another in terms of ideologies and practices. They are also the locations where resistance and acquiescence can occur simultaneously (Gaventa 1980; Sachs 1996; Seitz 1995). Let us turn now to a more contextualized version of this conceptual framework and how it applies to our understanding of the women's experiences underground.

GENDER, RACE, AND CLASS IN THE COALFIELDS OF COLONIZED APPALACHIA

Mainstream American society sees Appalachia as part of the "other America." Its material poverty and symbolic "backwardness" are juxtaposed against a more affluent and presumably more progressive culture. Geographically, although it consists of northern, central, and southern subregions, central Appalachia has epitomized these stereotypes and therefore has been studied by numerous scholars (see Billings and Tickamyer 1993).[5]

---

the differences in women's standpoints, it has been argued that such diversity does not deny women's common oppression and institutionalized subordination (Aptheker 1989, as discussed in Sachs 1996).

5. Since Appalachia's "discovery" by writers and scholars in the late 1870s, images and theories of the region have been plagued by stereotypes. To the early "local-color" writers, Appalachia was a "strange land" inhabited by "peculiar people" (Shapiro 1978). In later decades Appalachians were depicted as either lawless, moonshine-brewing hillbillies or "docile diggers," both inhabiting a perpetually "retarded frontier." In a similar vein, writers on the Appalachian "culture of poverty" have interpreted the region's economic woes as a result of

According to many of them, central Appalachia's problems are the result of "its integration into the national economy for a narrow set of purposes: the extraction of low cost materials, power, and labor, and the provisions of a profitable market for consumer goods and services" (Whisnant 1980, 129). The unfortunate combination of the region's relative economic isolation and underdevelopment when the northern United States was industrializing made it vulnerable to rapid and uneven capitalist development.[6] As a result, Appalachians lost control over the region's natural resources and local institutions. Having suffered both economic and cultural exploitation at the hands of external elites, generations of central Appalachians bear the scars of real poverty and the stigma of perceived inferiority. As Cable (1993, 69) aptly points out, "what is significant about Appalachians is their history of systematic, routine oppression."

In his classic work *Night Comes to the Cumberlands* (1962), Harry Caudill described the Appalachian region, in its economic development by gas, oil, timber, and coal interests, as "the last unchallenged stronghold of Western colonialism." He argued that "the nation siphoned off . . . its resources while returning little of lasting value" (325). Later, Lewis and Knipe (1978) made a compelling case for considering central Appalachia an internal colony exploited for its coal reserves.[7] The "exploitation

---

Appalachians' cultural characteristics of fatalism, traditionalism, and individualism (Weller 1965). But in the late 1960s and 1970s, in reaction to these stereotypes, a host of scholars set out to reconceptualize Appalachia, particularly central Appalachia, as a living example of the negative economic and social consequences of capitalist development (Billings and Blee 2000; Caudill 1962; Eller 1982; Gaventa 1980; Lewis et al. 1978; Whisnant 1980).

6. According to economist Paul Salstrom (1994), precapitalist Appalachia's independent economy featured a low supply of indigenous capital that prohibited the growth of small businesses. There is also evidence that the federal government restricted the flow of cash that could have supported local and regional economic development. The lack of capital was accompanied by population growth, soil depletion, and a shortage of tillable land (Billings et al. 1986; Salstrom 1994). Taken together, these factors made the region easy prey for outside corporate interests, whereby "Appalachia experienced the sudden penetration of mature corporate capitalist institutions rather than slow evolution of work relations," which "contributed to development of an underemployed labor pool that could be utilized cheaply in railroad, timber, and mining industries" (Billings et al. 1986, 161). This capitalist transformation turned subsistence farmers into coal miners living in company towns, marking the beginning of a two-class system of nonunionized mine workers and mine-owning capitalists (Eller 1982; Maggard 1994).

7. For this characterization of a colony and its inhabitants Appalachian scholars are indebted to Robert Blauner (1969), who developed a model of classical colonialism and applied it to exploited Third World regions globally. Blauner defines classical colonialism as "domination over a geographically external political unit, most often inhabited by people of a different race and culture; when this domination is political and economic, the colony exists subordi-

model," as it is also known, is characterized by the acquisition of natural resources and the exploitation of labor to benefit absentee corporate owners, as well as the destruction of local culture and outside institutional and administrative control of the region's indigenous populace.

Lewis and Knipe (1978, 24) assert that "it cannot be disputed that the coal interests came into the region 'uninvited,' that cultural patterns changed as a result of this intrusion, and that the area is controlled by representatives of the industry." Although they agree that fatalism and passivity are cultural traits of Appalachians, they argue that these traits are the result of Appalachians' adaptation to their own sense of powerlessness. In addition to acquiescence, however, central Appalachians are also capable of resistance to these outside forces (Gaventa 1980). These scholars further argue that a "condition of racism" is evidenced by the stereotyping of Appalachians as lazy and backward, even by the small, independent coal operators who constitute a separate class of local elites. Thus, two basic social classes and their mutually antagonistic relationship evolved from the oppressive practices of the coal industry in colonizing the region.

nate to and dependent upon the mother country" (395). Colonization is also an oppressive process operating within "a system of dominant and subordinate group relationships." This process begins with the colonizers' forced entry and is maintained by cultural and institutional beliefs about the colonized group's presumed inferiority. Technological superiority is the main source of the colonizers' dominance. The colonizers enhance this form of superiority by harnessing the natural and human resources of the colony, which in turn enhances their power over the colonized natives.

In the case of coal mining, corporate capital's technological superiority is accompanied by organizational changes in underground work itself. Advances in mining technology meant increasing mechanization accompanied by organizational changes, which allowed capitalists to gain control of the mining labor force. In most mines today, the process of getting coal is fully mechanized and continuous from seam to tipple. By increasing mechanization, the capitalist operators hastened the pace of production and increased the dependability of the output. To some extent they also successfully replaced labor with capital, increasing productivity. For miners not so displaced, increasing mechanization and the establishment of a job hierarchy in the mines did more than simply reduce their skill level. These developments transformed the miners' skills to better serve the operators' goal of accumulating more and more capital. (For a thorough Marxist interpretation of the coal industry in Appalachia, see Howes and Markusen 1981.) The establishment of a job hierarchy based on the recognition of different skill levels resulted in a corresponding wage structure. This enabled the capitalist operators to lower the average wage of miners and thus lower the cost of producing coal. But the "de-skilling" of miners had yet another effect that worked to the capitalists' advantage. In general, the required training time for becoming a miner was now greatly reduced. Hence, the removal of the skill barriers to entry-level mining jobs increased the competition for mining employment by expanding the pool of potential miners, creating what Marx termed "a reserve army of labor."

For more than a century the region's coal operators have employed institutional and cultural strategies to tighten their control over the miners. In his monumental work *Miners, Millhands, and Mountaineers* (1982), Eller shows that life in the coal camps from the late 1800s to the 1930s stood in sharp contrast to life and work on subsistence family farms.[8] Capitalists' rapid industrialization of central Appalachia relative to the northern and southern subregions allowed for the development of coal camp enclaves. Coal companies provided and controlled the most important "stabilizing" institutions in these enclaves, such as churches, schools, stores, and local law enforcement. Miners were paid in credit or in company scrip, not cash, which could only be used at the company store. Most of the time, lawmen were little more than security guards employed by the company. Company housing, which ranged from shanties to slightly sturdier structures, was used to supervise miners off the job and discipline them when they "stepped out of line." The coal operators also provided theaters, clubs, and various recreational facilities. In this way they were able to foster competition between coal-mining communities through sports, such as baseball. These paternalistic forms of control prevented miners from having a voice in community affairs and guaranteed virtually no interference from outside social welfare agencies.

Beyond the establishment of isolated company towns, coal operators hired diverse racial and ethnic groups meant to divide the labor force, prevent unionization, and provide a stable supply of mining labor (Ayers 1992; Caudill 1962; Eller 1982; Lewis 1987; Maggard 1999; Trotter 1990; Turner and Cabbell 1985; Shifflett 1991). Like white Appalachian men who were drawn into mining, African Americans from the South and immigrants from eastern and central Europe were also recruited into mining by coal company agents with promises of transportation,

8. Preindustrial Appalachia consisted of scattered, self-sufficient, and loosely integrated farming communities, not unlike many other rural areas in the South and Midwest (Eller 1982). The region itself was more or less isolated from the urban and industrialized centers outside it, and communities within the region were separated by the mountainous terrain. Even so, communities had a rudimentary communication system and minimal economic exchange between them and the outside world. Generally, though, these communities were islands unto themselves, and they demonstrated an internal stability based on their localized institutions and way of life. The family and extended kinship network formed the basic social, political, and economic unit. Land and attachment to place cemented Appalachians' identity, shaping their culture and social structure. Social stratification was based on minimal differences in political clout and land and livestock ownership, not occupation, education, or income. That would come later.

company housing, and a relatively high wage during the early 1900s.[9] Generally, race was more divisive than ethnicity (Maggard 1999). Although the highest level of racial mistrust existed between African American and immigrant miners owing to the cultural divide between them, racist attitudes toward African American miners, particularly among members of management, resulted in strict segregation in the coal camps (Eller 1982; Maggard 1999; Shifflett 1991). African American miners were also given the most arduous jobs and those with the least amount of authority underground (Lewis 1987; Trotter 1990). Given this racial segregation, Maggard (1999, 187) has referred to these African Americans as a "marginalized subclass within Appalachia's working class." Coal operators played black and white miners against each other in order to forestall unionization, and often brought blacks in as strikebreakers, which elevated racial tension in company towns. At the same time, white miners demonstrated racial prejudice, and lynchings were not uncommon (Eller 1982). In sum, the coal operators created a culturally divided, institutionally captive labor force that was dependent on, and to some degree powerless against, their domination.

The colonization of central Appalachia not only created two distinct economic classes further divided by race and ethnicity, it also established a division of labor within the family and sex segregation in the workplace (Maggard 1994, 1999). Although family life on the subsistence farm was patriarchal, women and children were integrally involved in all phases of household and farm production (Eller 1982; Pudup 1990). During the industrialization of central Appalachia, women played a more indirect role in the region's economy via their "'auxiliary' work of managing households and caring for dependents and the disabled" (Maggard 1994, 15). Many still tended gardens and livestock, but on a smaller scale than before (Lewis 1989; Pudup 1990; Trotter 1990). Some scholars contend that as the coal industry's domination of the region increased in the 1900s, women became more subordinate and dependent on men (Eller 1982; Maggard 1994). Just as men lost control

9. The African American population of Appalachia has never been very large, although there were some relatively large pockets of African American settlements in the region after migration from the South in the early twentieth century. In 1920, according to Eller (1982), 43 percent of African American miners in the United States mined coal in West Virginia, particularly in the southern part of the state. McDowell County, adjacent to the county in which the mine I studied is located, had the largest concentration. Both of the African American women in my sample came from coal-mining families, lived in McDowell County, and commuted to the mine.

over the economic means of production in the coal camps, women lost a measure of control over the means of household production, which increased their dependence on their husbands' mining wages. Many women began to participate in the "informal economy" by taking in boarders and doing others' laundry. The only other paid employment available to them was in nursing or as domestic servants and seamstresses (Maggard 1994; Oberhauser and Turnage 1999). Coal-mining families thus became more patriarchal than before; they became the "little white man's colony" in which working-class men could dominate their unpaid laboring wives and children (Mies 1986, 110; Seitz 1995, 93).

Along with the shift in women's work roles during the region's industrialization, Maggard (1994, 16) points out, a corresponding "two-pronged, gendered ideology emerged to justify this division and unequal valuing of labor." Breadwinning men did "men's work" in underground coal mines, while women's legitimate place was defined by their less valued, unpaid domestic labor at home doing "women's work," a condition Mies (1986) refers to as "housewifization." Not only did this set of beliefs about women and men correspond to the principles of patriarchal dominance, but this emergent gender ideology also served the interests of capitalists (Barry 2001; Eviota 1992; Gibson-Graham 1996; Woods 1995). Capitalists need women's unpaid labor in the home and reproduction of laborers in order to minimize their own costs. Keeping their costs low allows them to expand their operations and continue to accumulate capital. As Eviota (1992, 15) explains, women's work in the home "takes place outside the market sphere but is integrated into it: no worker, male or female, would be able to sell labour power to a capitalist unless basic needs were first met. In reality then, the household and maintenance work of women is the material base for the reproduction of living labour without which capital cannot appropriate surplus."

Because the housewifization of women supports both patriarchy and capitalism, it has at least two debilitating effects on women relevant to my discussion of women miners. First, when women engage in paid employment outside the home among men, they are subjected to both informal and formal pressures to perform the same duties expected of them as housewives at home. The result is a gendered division of labor that puts women at a disadvantage in the workplace. Second, as Seitz (1995, 28) has pointed out, this housewifization deters either "material or ideological connections across the borders of class, race, and political geography" between women that would serve their collective interests. As a result of these divisions, women are hampered in attempting to develop solidarity.

Historically, women's roles have been sharply circumscribed by rural Appalachians' traditional gender ideology (Maggard 1994, 1999; Pudup 1990). Even today, rural women's roles and corresponding beliefs about "a woman's place" are notoriously resilient (Tickamyer and Henderson 2003). Traditional notions about men's and women's work are strictly defined and ultimately reinforced by the interests of capitalist economic forces in the region. According to Ollenberger and Moore (1998), gender roles are particularly rigid in rural areas owing to the colonization process, whereby the expropriation of material resources and human labor more severely affects the status of minorities. "Under the colonization model, the control of value and of symbols is attached to race and cultural dominance through capitalism or other forms of economic oppression. The ethnocentrism of this system is developed and maintained through educational, political, and economic systems controlled by the dominant group" (Ollenberger and Moore 1998, 43). Specifically, women and blacks are symbolically represented as less skilled and inferior workers, and their corresponding exchange value as workers is set by men relative to male workers based on the "bureaucratic administration of labor." Thus, under the circumstances of colonial industrialization, the ideologies of both sexism and racism receive institutional support.

Within most industrialized nations, socialist feminists have focused on the generally mutually reinforcing (though occasionally conflicting) relationship between capitalism and patriarchy as institutionally manifested. Hartmann (1976, 138), has defined patriarchy as "a set of social relations which has a material base and in which there are hierarchical relations between men, and solidarity among them, which enable them to control women." Under the system of patriarchy, male privilege is preserved via the oppression of women, just as under the system of capitalism, capitalists' privilege is preserved via the oppression of workers. With the state's alliance with the capitalist economic system, however, the accompanying loosening of private-public boundaries threatened working-class men with the erosion of their privilege within the family. That is, by employing women outside the home, the capitalist system weakened men's patriarchal authority over women's labor power within the household. Using the hierarchal organization and control they had mastered under the patriarchal system, men maintained their privilege by reproducing their control over women both at home and in the workplace. A domestic division of labor emerged whereby male miners controlled and benefited from their wives' unpaid labor at home, while women in the workplace were relegated to lower-paying jobs within male-dominated occupations. Either way, women are kept dependent on men as men's patriarchal control over them tightens under the patriarchal capitalist system.

Hartmann adds that "patriarchy is not simply hierarchical organization, but hierarchical organization in which particular people fill particular places" (1984, 180). Since patriarchal relations are reproduced in the workplace, socialist feminists emphasize "the role of men as capitalists in creating hierarchies in the production process in order to maintain their power. Capitalists do this by segmenting the labor market (along race, sex, and ethnic lines among others) and playing workers off against one another" (Hartmann 1976, 139). Under these circumstances, men are united by their common interests in maintaining the status quo and are therefore dependent upon one another to make these hierarchies work. Men at higher levels in the hierarchy "buy off" those at lower levels by offering them power over individuals who are even lower in rank. This is how women become exploited by capitalists as workers, but also by other men, resulting in their "super-exploitation." This is also why Hartmann argues that male workers and their unions, in addition to male supervisors, help maintain the sexual divisions found in the workplace. Therefore, segregation of work by sex is the primary mechanism by which men preserve their superiority over women.[10]

As women began working in underground coal mines, they faced men who attempted to label and control them both as women and as workers. As women, they frequently faced men's hostility and sexual harassment. As workers, they were victims of supervisors' selective use of promotion policies and other forms of sex discrimination. Even though the company complied with the state's mandate to hire women,

10. A related theory appropriate to this investigation is "social closure theory," which holds that "a status group creates and preserves its identity and advantages by reserving certain opportunities for members of the group" through exclusionary and discriminatory practices (Tomaskovic-Devey 1993, 61). Because women pose a threat to men's privileges, men will tend to emphasize women's presumed illegitimacy for doing masculine-identified work. They preserve the gendered status hierarchy through "certain social practices that create or exaggerate the social distance between status groups" (Reskin and Roos 1987, 7). These practices dictate subordinates' behavior in the presence of members of the dominant group and shape the casual interactions between them.

male managers swiftly realigned their interests with male miners by making job assignments based on sex. In this way they allowed male miners to "choose" their jobs so that they could continue to maximize their economic advantages in the workplace (Strober 1984). As in other male-dominated occupations, once a sex-based division of labor has been established, it is extremely difficult to change (Milkman 1987). The message was clear: if women were going to work underground, they would support and serve men. In short, the women were expected to do jobs that benefited men underground just as they were expected to do as wives aboveground. Their gender-based assignments to lower-wage, less skilled jobs also held them back from effectively competing with men for more skilled, better-paying ones (Sokoloff 1988, 128). Nonetheless, as we shall see, despite what sometimes felt like overwhelming odds, these women persevered by resisting either individually or, on a few occasions, collectively men's attempts to "keep them in their place."

THE BOOK

My main aim in this book is to give the reader an in-depth view of the working lives of central Appalachian women miners by listening to the women's voices. As you might expect, coal mines are dark, damp, and often eerie places where miners usually work side by side in cramped areas. Their work is extremely dangerous and they must depend heavily on each other. Coal mining also continues to be an occupation with a strong masculine-identified subculture, and women miners face formidable barriers to their full integration into the workforce. Before I began my study, I talked informally with women miners from across the nation about their work experiences with men. I also read reports in which women miners confirmed that the buddy system underground extends beyond work relations (Dawson 1990; White, Angle, and Moore 1981). Although one might expect supervisors to use women employees to undermine the "brotherhood" engendered by union solidarity, male miners and supervisors frequently became partners in patriarchal control underground (Yarrow 1985). As a result, many women experienced sexual harassment and more subtle forms of "sexualized work relations" at the hands of some male miners and supervisors or "bosses." (For further discussion of the sexualization of work relations, see Enarson 1984.) They had difficulty advancing to highly skilled jobs. I also learned that union leaders reacted negatively to women joining the rank and file. But women had slowly begun to make themselves known by attending meetings locally and internationally and by serving on mine and safety committees.

Women's on-the-job experiences with male miners and bosses, their attempts to advance, and their relationship with their union at a large coal mine in southern West Virginia constitute the major focal points of this book. I found that the women's accounts were as empowering for them to tell as they were for me to hear. Because they had entered the male-dominated and masculine-identified world of underground coal mining years before, they had developed particular points of view toward themselves, the male miners and bosses with whom they worked, family members, and members of their own mining communities. What could these women tell me about those who subjugated them? How did they negotiate their identities as women while working in a masculine-identified occupation and, more tangentially, while living with their families and in their communities? What were their modes of accommodation and strategies for resistance? What threatened their solidarity with one another? What common knowledge brought them together as miners and as women?

To answer these questions, I examined the work experiences of a single cohort of white and black Appalachian women working underground over an extended period of time. What I learned from my interviews and conversations with women, men, some managerial personnel, and members of the local community is contained in the following four chapters and the epilogue. Chapter 1 focuses primarily on their early years of employment and the men's attempts to sexualize work relations, while examining the women's strategies for accommodation and resistance. Chapter 2 covers the women's ways of adapting to their work and their advancement experiences in light of the established sexual division of labor underground. Here I emphasize how informal social relations between women and men inform the use of the formal practices that govern promotion. In Chapter 3 I discuss the women's participation in and experiences with their union, the UMWA, mainly at the local level, while examining the gendered elements of women's and men's class consciousness. Chapter 4 describes the women's socialization as miners and examines the bonds and divisions that developed between them over the years. The epilogue discusses the extent to which these women changed men's attitudes about them and the general outlook on an industry in decline.

1

DIGGING IN: COPING WITH SEXUALIZED WORK RELATIONS

In the 1970s, Appalachian coalfield residents experienced both the prosperity of the coal industry's economic growth and the cultural changes wrought by the social turbulence of the previous decade. Beginning with the oil crisis in 1973, the industry enjoyed a boom that lasted into the early 1980s. Coal-mining companies began hiring underground miners in numbers unparalleled by the previous boom. The clarion call among the region's underground coal miners was "Come blister and bleed." This new generation of miners included veterans back from the Vietnam War, black Americans invigorated by the civil rights movement, and women inspired by another wave of feminism. They were younger, better educated, and more politically astute than the previous generation of miners (Perry 1984). They were also more likely to question authority. The mine superintendent, a rather intense, well-educated man who had once been a miner, said miners went out on wildcat strikes over virtually anything.[1] According to one veteran male miner, "In the seventies, we

1. Like the 1930s, the 1970s was a period of conflict between operators and miners, which affected the terms of miners' employment significantly (Simon 1983). As a result of the energy crisis during the early part of that decade, both parties anticipated rapid growth within the industry. The miners had high expectations for winning numerous concessions in the 1974 contract, and to some extent their expectations were fulfilled. The contract mandated monthly meetings of the mine committees, which were composed of miner and management representatives, and streamlined the grievance procedure. Although the companies continued to look for ways to cut costs, they also agreed to increase miners' wages and benefits. This was done to attract new miners and increase production, a move that made mining more attractive to nontraditional employees such as women.

could basically do what we wanted to do and get away with it." But the restlessness of the rank and file was also attributed to their dissatisfaction over more serious issues, such as health benefits (Moore 1966b). At the same time, there was considerable turmoil among the UMWA leadership, as Arnold Miller's "Miners for Democracy" unseated the UMWA's corrupt and entrenched president, Tony Boyle. Only during a time such as this could women have been tolerated, and the company began to hire them in record numbers, thanks in large part to the Coal Employment Project's successful class action suit.

GETTING HIRED

Between 1973 and 1977 women in the central Appalachian coalfields were hired on their own, one at a time. Often they had to file complaints with state agencies charged with enforcing equal employment statutes before they could get hired. In 1973, using individual women's complaints, the Kentucky Human Rights Commission began targeting the state's coal companies; as a result, the number of female coal miners in the state jumped from five in 1974 to two hundred in 1977 (*Mountain Life and Work* 1978b). But according to advocates of women in mining, such agencies were slow to recognize the obvious and broader discrimination in the industry (Hall 1984; Thrasher 1981). This became more evident when women belonging to two grassroots organizations in the region were denied an underground mine tour. The mine operator, also

---

Over the next three years, however, miners unexpectedly went on a record number of wildcat strikes. They felt their safety had been seriously compromised as the operators stepped up production. Moreover, the anticipated growth failed to materialize and the industry was generally regarded as being in decline. The disappointments experienced by both parties were reflected in the struggles over the 1978 and 1981 contracts. In the former agreement, miners lost certain health and welfare benefits along with cost-of-living increases. In the 1981 contract they suffered even more "take-backs," including limitations on their right to bid on jobs. Conflicts between labor and management persisted throughout the 1980s and into the 1990s over such issues as mine and machine safety, the flow of mine communications, union jurisdiction, job-bidding rights, and the handling of miners' grievances. Miners felt betrayed by their union leaders and believed that without change in those representing them, even further concessions on their part would follow in the next contract. With the election of Richard Trumka in 1982, miners entered an era of renewed militancy and relatively successful attempts to regain the contractual losses of the 1970s, if not the jobs lost to increasing automation and the recessionary pressures of the early 1980s. It was during this period that many recently hired miners, particularly women, were laid off in staggering numbers.

a criminal court judge in eastern Tennessee, told them that he couldn't "have no women going underground. The men would walk out; the mine would shut down. . . . If you insist on bringing [them], forget the whole thing" (Thrasher 1981, 48). This led to an investigation by attorney Betty Jean Hall, who released to the media her finding that 99.8 percent of all coal miners were men (Appalachian Alliance 1982; *Mountain Life and Work* 1979, 1984; Thrasher 1981). After touring the coalfields herself, Hall was overwhelmed by the number of women who said they would take a mining job if they could get one. With entry-level pay beginning at about $60 per day under UMWA contract, Hall and other women's employment advocates saw this as one way to alleviate poverty among central Appalachian women (*Mountain Life and Work* 1978b).

In May 1978 newly installed Coal Employment Project director Hall and her staff filed a complaint against 153 coal companies. The basis for the complaint was Executive Order 11246, which obligated coal companies that had contracts with the federal government to follow its fair employment practices and affirmative action guidelines. By filing a complaint through a government agency, in this case the Department of Labor's Office of Federal Contract Compliance Programs (OFCCP), the CEP could get the federal government to act on behalf of women seeking jobs underground. Six months later, some of the nation's largest coal producers settled the lawsuit by paying thousands of dollars in back pay to women who had been denied jobs and agreeing to hire women until they constituted approximately one-third of the total workforce (Appalachian Alliance 1982; *Mountain Life and Work* 1979). The CEP also issued press releases informing women about coal-mining jobs and encouraging them to apply by initiating their own lawsuits. During the following summer and fall, the CEP received dozens of calls and letters from women seeking mining jobs (Hall 1984). When the CEP was created in 1977, women made up just 1 percent of the total coal-mining workforce (*Mountain Life and Work* 1979). In 1978, they made up slightly less than 5 percent of all new hires in the industry. By 1979, that figure had increased to 11.4 percent, as their absolute numbers began to rise rapidly (Appalachian Alliance 1982; Reskin and Hartmann 1986).[2]

2. The Coal Employment Project helped women get mining jobs by bringing legal action against the coal companies, organizing various recruiting efforts, and sponsoring training programs and support networks for women workers from Appalachian coal communities. Being isolated and scattered around the nation's coalfields, these women had little if any contact with one another. As the CEP and its mission became more publicized, coal-mining

Like other coal companies, the company I studied did not begin hiring women in appreciable numbers until it was forced to do so. Before the 1978 lawsuit, the company employed five female miners. Hired in 1975, Sadie was one of the few women "pioneers" still working at the mine.[3] As a result of her years on the job and her caring nature, she was often identified as a mother figure by many of the younger women miners. According to Sadie, women who had applied for jobs underground before the lawsuit were routinely turned down, because the company could hire women at its own discretion. She suspected that, at the time, the company "wanted to hire mostly married women who were more settled," adding, "when we went in, the miners' wives didn't want no women in the mines. They still don't like it. Most men don't want women in the mines," either. In September 1978, Jolene, a savvy and physically attractive woman then in her early thirties, had applied for a mining job with the company and been turned down. "My ex-husband was working here is the only reason I figured they didn't want to hire me." But she persevered. "After I first applied, I came back for several weeks every week," she said. "Then I just totally forgot about it. Later I ran into two women who had been hired. I had just a day or two left on my underground card and I would have to take the [mining] class over. So I called and asked why they hadn't hired me. They said they just weren't interested, and I said I knew they had hired a lot of women and that sounded like an excuse to me. They said it's up to you if you want

women from around the country began to contact the organization to find out about other women miners in their areas and how they could organize themselves locally and regionally to solve their problems and support one another (Hall 1984). Eventually, a small number of active groups of women miners banded together as a part of the CEP under an umbrella organization called the Coal Mining Women's Support Team (CMWST). In 1979 the CMWST began publishing a quarterly newsletter and, along with its sister organization, the CEP, cosponsored annual women coal miners' conferences, held throughout the country, for the next twenty years. But as the coal industry lapsed into decline and employment figures plummeted, so did the number of women miners, per the "last hired, first fired" rule. As a result, the CEP faced financial problems of its own. With much regret, director Cosby Totten, a laid-off miner herself, and her board of directors shut down the CEP in July 1999.

3. Walshok (1981) first coined the term "pioneer" to refer to the first women hired in nontraditional occupations through their own initiative, which may have included filing individual lawsuits with state or federal agencies. As one would expect, these women tend to have more difficulty in the workplace, bearing the brunt of men's hostility more often than those women hired later under affirmative action agreements. Although the pioneers generally see themselves and are seen by others as being tough, confident, and willing to take a chance, they also tend to have higher quit rates. For further discussion of women in nontraditional occupations, see O'Farrell (1988). Most of the women in the study were hired during the postpioneer period.

to push it. So I called a man who knew the woman who was handling that suit, and the next morning they hired me."

In 1977, Michelle, a stout black woman then in her midtwenties who was reputed for being blunt and outspoken, had also applied for a job with the company and been turned away, which she attributed to racial discrimination.

> I put in an application and I went there every day. I sat around the office all day long, every day. Back then, that would show the superintendent that you really wanted the job, and [if you did] it long enough, [he would] eventually hire you. That superintendent did everything but just about slap me in the face. So one day I said to myself, I don't have to take this, and I did not go back for over a year. They had hired four to five white women and during that time they did not hire any blacks. Some of my neighbors knew I had put in an application for it, so they called me that day. My course had expired, so I did have to take the mining course over again. I met a friend of mine, and we took the mining course together, and we both passed. The same day that I went up there she went, too, to put in her application. She was hired on the spot right there with me.

By the fall of 1978 the company, along with several others operating in the central Appalachian region, was being sued by the Department of Labor for sexual discrimination in hiring. Following an investigation by the OFCCP, the company settled the suit by paying back wages to those women it had turned away and by adopting a new hiring ratio beginning in 1979. Hired in 1978, Carolyn was a divorced mother of several young children in her early thirties. She recalled that during the summer of 1979, "There were about five to six women there before I went in. [That's] when the other women sued [the coal company]. They were needing to hire women in sixty days. Some of them got a lot of money back from when they put their application in. I didn't get any money and Kelly didn't get any either, but she was hired in something like thirty days." "At one time," said an older male miner, "we had a lot of women apply for jobs and [the company] wouldn't hire them. Then [the company] hired a gang of men. It went on for seven or eight years. Then the women put in lawsuits. They collected big money. All of them did." One of those women was Jolene. "I filed that claim and I got that money, but I would have more seniority than between five and ten other women if they had hired me then." Even so, to many people's minds, the loss of seniority was no match for the thousands of dollars these women did collect, because, as miners of both sexes agreed, the money only added to some men's resentment as these women began working underground.

The superintendent, who was the personnel officer at the time of the settlement, explained that the company operated out of fear. Because of pressure from the federal government, he said, he was unable to screen female applicants as he would have male applicants, and the women knew it. He remarked that their attitude was, "You have to hire me; the government said so." So the company was forced to accept virtually any female employee who applied. In dealing with women applicants, he said, the company "backed off even when it was technically, legally, and morally right." He expressed resentment at the government for infringing on his right to manage the workforce. "I gave up a right when a woman got something [a coal job] that she didn't deserve," he said. "Management had to pay the price for social change." In his view, none of the women hired was suitable for a mining job. Moreover, he said, this wholesale entry of women into the mine required changes in company policy to accommodate the relatively few but highly visible new members of its underground workforce. During this period the company took steps to ensure that its newly certified women miners were trained on equipment, and it established rules governing miners' conduct underground. These changes were made in response to lawsuits filed by women at other mines over discrimination in advancement and sexual harassment.

As I was to realize later, the superintendent's comments reflected his commitment to the prevailing gender ideology of the patriarchal system and his role as an agent within the capitalist system (Scott 1995; Yarrow 1985). As Scott points out in *Two Sides to Everything* (1995), hiring women would benefit capitalists, both because they could pay them less and because they could use them to fracture the male solidarity of the union. In fact, however, the male coal operators Scott studied aligned themselves with male miners by allowing them to sexually harass women and permitting a gendered division of labor. This was not formally the case at the mine in my study. However, particularly during most women's first few years underground, bosses, being out of sight of their supervisors, often did side with men miners, to a woman's detriment.

For the superintendent at the mine I studied, the advent of women miners meant a loss of managerial control and, to a greater extent, the erosion of male privilege. Echoing racist rhetoric about miscegenation, he added that the mixing of the sexes underground would cause irreparable turmoil. He believed that most of the women hired as a result of the lawsuit were unfit and not the least bit serious about being coal miners. As he put it, many of them took mining jobs because they were "looking for adventure or for a sugar daddy. Some were straight-laced and some were sluts, just like some of the men," he said. Similarly, one of the male miners, a man who began working underground in the 1960s, declared that, at the time, the "majority of [the women] just didn't give a damn. They just sort of played at whatever they had to do, they really didn't want to work." He also referred to most of the women as "gold diggers." Men's charges of sexual promiscuity were corroborated by several women miners who said that during their first few years there had been numerous affairs between women and men miners and bosses. Sally, for example, a candid, sincere, hardworking woman who began mining when she was thirty, had rebuffed several men's advances. She declared, "At one time there, I bet, over half of them [women miners] was sleeping around, married, not married—didn't matter."

As in previous studies of women miners, however (Hall 1990; Hammond and Mahoney 1983; Scott 1995; Yount 1986), the women I interviewed said they needed coal jobs, which paid better than the other jobs available to them, primarily in order to support either themselves or their families or both. Often they were sole breadwinners providing for several children or, in a few cases, grandchildren. They had decided to get mining jobs because they were divorced or because their husbands had become disabled or been laid off from their jobs. Before being hired at the mine, many of them had held lower-paying jobs traditionally held by women, such as working at a bakery, clerking in a store, or selling hot dogs at a drive-in, jobs that barely enabled them to get by. The income from their coal-mining jobs afforded them a more desirable middle-class existence and thus a sense of self-sufficiency, independence, and empowerment that some of them had never known. As Carla, a tall, attractive, vibrant, quick-tempered woman in her late thirties, recalled with considerable bitterness, "When I was growing up we didn't have nothing. We had hand-me-downs. We lived on welfare. Mommy raised four kids by herself. Daddy and Mommy divorced when we were little. We always knew Daddy, but we lived with Mommy. Kids made fun of us in school. They called us all kinds of names and I'll never forget it. I always said, 'When I grow up, I'm going to get a job and people are going to stop making fun of me.' Like I told [my husband] when he was trying to get me to quit, 'I will not quit and bring my kid up like I was brought up.'"

Reflecting upon her upbringing, Renee, a short, soft-spoken black woman in her midthirties, remarked matter-of-factly, "I was raised up on welfare and I wanted better for myself when I got out of school. The opportunity came along and I knew I had better take that job. My husband was out of work. I wanted better for myself, for me. He's off, and there are things that I want. Then, if I had kids, I wanted to be able to give to them what I never had. Even though I had a husband who could do it, I wanted to do that. It's something that you have of your own."

For both women, having a coal miner's income meant that they had finally escaped the shame of being raised on welfare. Being Appalachian carries a social stigma similar to the stigma that comes with being black; both groups have been identified as "cultures of poverty," their impoverishment resulting from their own "internal deficiencies" (Seitz 1995). Both white and black Appalachian women were also similarly affected by the controlling image of the "welfare mama," and getting work in the mine represented their own individual resistance to it.[4] The ability to do a traditionally male job requiring physical stamina and courage was an added reward, and for most of the women I interviewed, being a coal miner was also a source of pride. Kelly, a short, athletic woman who started mining in her early twenties, was single with no children. Along with a tremendous work ethic, she had held one of the most physically demanding jobs longer than most other women miners. Toward the end of our second interview, she said,

4. Underground, the women confronted and resisted racial and gendered stereotypes in the form of what Collins (1990) calls "controlling images," which serve to identify and define subordinate group members in categorical terms by objectifying them as inferior "others" and denying them their individuality as well as their common humanity. Both Maggard (1998) and Collins (1990) have specifically referred to white Appalachian women's "Daisy Mae" and "Granny Clampett" stereotypes and black women's mammy and matriarch images, respectively. These controlling images for white and black Appalachian women become their badges of inferiority as workers. While white women are considered to lack the ability to do a "man's job," black women experience the double stigma of inability and indolence because of their race.

> Being a woman and being in the mines and being able to perform your job as well as a man? Yes, of course, I'm very proud of that, and I work hard every day. When I go places, people I don't even know will ask me, "What kind of work do you do?" Just by looking at me. It looks like you do physical labor work, whatever. It is never negative when you tell them. They are always intrigued, they always ask questions.

Being able to support themselves and the children they had was a point of pride for these women, even if it meant challenging regional working-class beliefs about women and work.

At its peak in 1980, the company employed about eight hundred miners, of whom ninety-seven were women. But over the years their numbers had dwindled to one-quarter of that number. Some of the women hired as a result of the lawsuit were eventually fired, while others married and quit. "There have been a lot to come and go," said Irene, a stocky, plainspoken woman in her midfifties, now retired after sixteen years of mining. "Some just didn't want to work, some just got married and quit, and some just moved on," she said. Moreover, during the mid- to late 1980s, improvements in mining technology and industrial decline forced even the largest coal companies to lay off their least-senior miners, including many of the women. The superintendent said the layoffs gave the company "the excuse it needed to get rid of those women who weren't qualified to be coal miners." But he did allow that the women who had stayed were the "cream of the crop. We don't even think in terms of male and female anymore." But because, years ago, the company was forced to hire practically all its female applicants, he said, those women who were serious about their jobs suffered a backlash of negative sanctions from the men at the mine and in their communities. Indeed, the lawsuit had brought women miners almost as much unwanted attention aboveground as it had created tension between women and men working below it.

ENCOURAGEMENT AND OPPOSITION IN THEIR HOMES AND COMMUNITIES

The miners I studied lived in small, quiet, relatively isolated working-class towns nestled in the hills and hollows typical of central Appalachia. Members of the local community were bound almost exclusively to job and family, and the social worlds of work and home were essentially one and the same. Social relations between mining and nonmining community members were based on strong ties to family and extended kinship networks, which were often reproduced at work. Although women's and men's worlds were not entirely separate, support for traditional gender roles remained strong, as in other Appalachian coal-mining communities (Maggard 1999; Scott 1995; Seitz 1995). Women tended to be employed in service jobs, occasionally working alongside men, or else worked at home keeping house and caring for their children. In general, men were identified by their occupational roles as the family's principal breadwinners, while women were identified in relation to their husbands and families even if they worked outside the home. Much of my own interaction with local men and women revealed not only a fairly strict gender-based normative code of conduct, but also a double standard for men's and women's behavior. Like the men that Scott (1995) studied in a southeastern Kentucky mining community, the men I observed were afforded liberties, such as drinking, cussing, flirting, and being sexually "loose," without the stigma that would attach to women for the same behavior.

Previous research has shown that when it comes to women in mining, traditional gender roles have more or less persisted among members of Appalachian coal communities (Giesen 1995; Scott 1995; Seitz 1995; Trent and Stout-Wiegand 1987). Generally, these studies have found opposition from male miners and from nonmining members of the community to women working in the mines, particularly among miners' wives. However, different parties have different reasons for their opposition. In their survey of community resistance to women coal miners in a northern West Virginia coal-mining town, Trent and Stout-Wiegand (1987) found that working women and male miners were more likely to support a woman's right to a mining job than were housewives or nonmining men. Male miners were also more likely to encourage their daughters to be coal miners, because of the high wages. The researchers surmised that male miner's pro-union attitudes and working women's challenge to traditional gender roles made them more likely to support equal rights for women. But male miners were also more likely than women, regardless of their work status, to believe that women could not be as capable as men working underground and that coal-mining women could no longer be considered "ladies." Housewives were even more conservative on this issue because they were presumed to be miners' wives or sympathetic friends of miners' wives. These researchers concluded that a woman's decision to seek a coal-mining job was likely to be influenced by the opinion of other members of the tightly knit rural community, a consideration that is less a factor in urban settings. Similarly, Giesen's 1995 study of West Virginia miners' wives revealed that even though they generally supported a woman's right to work, especially if she had children to support, most of them were adamantly opposed to women working underground. These wives believed that women were not physically suited to mining coal, that they would create hazardous conditions underground, and that the men would treat them differently. Likewise, Scott found that many male miners in southeastern Kentucky believed that women were incapable of working underground and that it was culturally inappropriate for them to do so. According to the local gender ideology, women were defined as "weak, incompetent, and emotional," and members of the local community used ridicule and social stigma to show their disapproval (Scott 1995, 91).

My own findings about family and community reactions to women in mining echo many of these beliefs. A veteran male miner who was generally sympathetic to women miners said he had no problems working with women underground. He told me stories of women who had been more dependable and hardworking than some of the men. But, he said, "As a general rule, people believe that women don't need to be in the workplace anywhere, let alone a coal mine. Most people were negative against it, and probably those that wasn't were less than honest." Apparently Maggard's (1994) "two-pronged gendered ideology," which supports the stricter patriarchal division of labor found in the coal-mining regions of Appalachia, is still dominant.

The women I interviewed said that when they first started working at the mine, some of the opposition came from ex-mining male members of their families, but most came from nonmining males and miners' wives in the community. But the longer they had their jobs, the more their male coal-mining kin became accustomed to the idea. Kelly's grandfather, a former miner, "begged me not to go in. It was the old ways," she said. Her father, also a retired miner, had forbidden her to work in the mines. He declared, "No daughter of mine will ever go in the mines." But, she said, "I wasn't living at home at the time. I said okay. I was underground four months before he ever found out." By that time, "there was nothing he could do about it. He accepted it." These days, when she visits him, they talk for hours about mining. "We mine coal every time I see him," she said.

Lynn, then in her early thirties, was divorced with two young children to support when she began working underground. She said her father, a retired coal miner, was "the one who told me not to work in a restaurant when I could be making this kind of money to raise my kids, because my kids were three and four when I started in the mines." Other women said that their fathers and brothers have bragged about them.

While grandfathers, fathers, and brothers ultimately expressed pride in their female mining kin, boyfriends and husbands were more oppositional. Given the nature of the patriarchal family, boyfriends and husbands were affected differently by their coal-mining girlfriends and wives. When the woman mining coal was "their woman," her independence more directly threatened their claim to male privilege, power, and status within their own "colony." Darlene, an older, husky woman with a reputation for being feisty and outspoken, said her kin supported her decision to work underground and that she had "paved the way" for two female cousins who started mining after she did. But her live-in boyfriend left her and her three children after caving in to pressure from his men friends, who asked him why he couldn't "take care of his woman" so that she didn't have to work underground. In his absence she continued working, and a year later he returned home, telling his friends that she "had a mind of her own."

Jolene's experience demonstrates just how threatened some men could be by a woman's independence. "My ex-husband came up here one night after I started working here," she said. "He held me up against the fence and wouldn't let me go to my car. I had left him. He wouldn't work. He beat me. He about killed me. Even after I came to work here, he'd blacken my eye, thinking I wouldn't come to work [looking] that way. [But] I did." She explained bitterly, "He knew that I didn't have to depend on him any more. I had a good job and I didn't need him and his crap."

Children's reactions to their coal-mining mothers were mixed, even though these women had gone to work underground primarily to support them. Irene, now retired from coal mining, was in her midforties and divorced and had seven children when she started mining. One of her sons and two daughters had followed in her footsteps. Ironically, another son and daughter had objected to Irene's working underground. During our interview at her home, she and her daughter, a housewife, debated the issue of women in the mines. Irene had just finished explaining how women miners had been "put down a whole lot," when her homemaker daughter remarked that women didn't belong underground anyhow because mining was a man's job and women simply weren't as strong as men. "Oh, bull!" Irene shouted. "I handled it for sixteen years. It *ain't* no man's job." To which her daughter replied, "It will bring a woman down faster than it will a man." But when Irene reminded her that she raised all her children on a miner's pay, her daughter conceded the point, saying, "It's good money, and if they were hiring, I would probably do it because I have two kids to raise, too. But as far as the physical part, I don't think a woman can handle it like a man." Irene quipped, "Well, I wish she'd shut up," adding emphatically, "and it *is* good money." According to some women in the community, women could only appropriately do a "man's" job as long as they, too, had a "man's" financial responsibility to support a family.

The nonmining men in the community disapproved of women in the mines because they too felt that women weren't capable of doing the work. The men, they explained, would have to pick up the slack. But they routinely failed to consider that men also cover for less industrious men. Sadie recalled that a neighbor, a former miner, let her know how he felt. "I used to have long fingernails and I kept them pretty. I never had any problem. I just kept them cut off," she said. "Well, I had one guy ask me, 'Who's doing your work, the boss or the fire boss? Because there ain't no way you can work with fingernails like that.'" Another man in her community who had never been a miner told her she had no right to a mining job when so many local men were out of work. She responded that she had to make a living just like they did. As Michelle put it,

> I think a lot of men working outside the mines have a problem with those of us working inside the mines. I know this one guy [and] he says, "Well, you're a woman [taking a man's job]." I said, "I know a coal mine that opened up over here. Why don't you go over and put in an application?" [Later] I seen the same guy. So I asked him, "Did you go over there and put in an application?" No. He ain't even went. And this man has his electrician papers and stuff!

The women working at stores and restaurants remarked to me that if a woman could get a good job in the mines, she had a right to work there. I had told an older woman working at the diner where I ate regularly that I had gone underground that day. She said enthusiastically, "That's a woman's mine, you know. And if I was younger, that's where I'd be working." She said that as far as she was concerned, there was nothing inappropriate about women working underground. But Jolene, then in her midforties, had several working female friends her age who "think a woman's not here [in the mine] to do her part, she's just here for an easy ride." She attributed this attitude to sheer envy over her bigger paycheck.

> I went in a store one time, and a girl I went to school with worked in there. I had to cash my check. I had worked so many hours. [She said,] "I don't make as much in a month as what you make in two weeks." That sort of made me mad. At the time, you could get a job in the mines. I told her, "Go and apply. What's stopping you?"

In a similar episode, Michelle said,

> I went into the bank and I wanted to cash my check. The woman [working there] said, "I want you to sign the name of your husband [on the check]." I said, "Why? It's mine." It made me laugh. She said, "Are you a coal miner?" I said, "Yes, it's mine." Then they ask you questions like how can you stand to go down inside there. How can you stand to get so dirty? I say it's all in a day's work. I take a cake of soap and wash it off. That's how I can stand it.

Like other members of the coalfield community, some working women approved of women in mining only if they had children to support. But even then, women miners were culturally enigmatic. While they understood that women miners had children to support, some working women found it inconceivable that women would want to mine coal for any other reason.

With few exceptions, women miners were resented most deeply by miners' wives. Ellen, a sincere, affable woman in her midforties, was still emotionally shaken by what a male co-worker had once told her. "There was a woman [miner] killed, I believe a red hat, in one of the other mines. And this guy said that his wife said that she wished that every one of us [women] at [the mine] got killed. That really bothered me for a long time." Some of the women miners felt the wives were jealous of them simply because they "were out working, making a living and independent." Recalling the days when she too was "just a housewife," Ellen admitted, "It was worse than anything I ever done in my life. I was jealous of the world because I wasn't a part of it [like] I wanted to be." However, as Scott (1995) also discovered, miners' wives feared that women miners would take either their husbands' jobs or their husbands. "There are a lot of women who are afraid of women working with their husbands and, you know, some of them probably have reason to," Lynn said sheepishly, with a chuckle. But most of these affairs were in the past, she said. Indeed, the general consensus among men and women at the mine was that many of the women who readily had affairs no longer worked there. Ironically, one of the older men blamed divorces on the wives' own infidelity. He said that about half the men at the mine had divorced and remarried, but not because of the "trouble" the women miners were presumed to have caused. Rather, he said, "The problem's with [the company] and the miners' wives." The men "don't spend enough time with their wives. The wives have got plenty of money, but they don't have their husbands with them. They're at work all the time. So the women go out with other men and before you know it, oh man, we've got divorce after divorce." Owing to the unstable and sometimes rancorous relationship between coal operators and miners, Yarrow (1991, 304) contends that "miners may have less in common with their wives now than their parents did in the coal camps of the past," and that the wives' financial interest in having their husbands support the family impedes miners' collective resistance to coal operators' control.

But the wives' resentment and charges of promiscuity were contested by the women who worked underground. Jolene recalled with lingering animosity, "We were at a Labor Day celebration and I was with my husband. I saw this [miner's wife] just pass by me, and she started giving me real dirty looks, and I said, 'Do you have a problem?' She said this man was her husband, and we work with one another. [I said,] 'Lady, I've got a husband, and I wouldn't have yours if you gave him to me.'"

Another story I heard repeatedly concerned an incident at Carla's doctor's office, where "people get to talking about the coal mines or somebody knows you. You're sitting there answering their questions and you got women over here saying, 'I don't think women ought to be in there.

Women in there ain't nothing but trouble.'" She asked them, "If something happened to your husband, and he was out of work, and you had to go in there, would you want to be talked about like this? Would you want to be called names or would you run to the welfare, sign up, and draw food stamps and a check, and let us working people pay the taxes for the money that you're getting? I didn't get an answer back." The women miners were disgusted by the wives' allegations. "I have morals and I have standards," Kelly stammered. "If I loved a man and he had a wife and kids, he belongs to her and that kid. I don't care how much I loved him. I would not break up a family under any circumstance. I think that's where [miners' wives] need to step back and reevaluate themselves. Do they not trust their men enough to work with women in a dirty coal mine?"

But not all of the men's wives had negative attitudes toward the women. An older miner's wife said she was not bothered at all by women working with her husband. As it turned out, she had been best friends with Irene long before she started working underground in 1978, and nothing had changed between them. She believed that the women in the neighborhood were divided over this issue. Similarly, some of the women miners had very positive relationships with miners' wives. When she first started mining, Carolyn, a slight, soft-spoken woman said that although "some people told me I couldn't do it, a lot of women [who were miners' wives] have told me, 'I respect you. I couldn't do it.'" Renee said, "When I run into somebody who knows I work in the coal mines, they'll say, 'Do you know so-and-so? What's your name?' I tell her, and [she says,] 'My husband talks about you all the time.' One man's wife, we talk to each other as if we've known each other all our lives." Indeed, Sally, her twelve-year mining career cut short by a back injury, thought that the men were more to blame than their wives, because even when the men's "wives were real sweet, a lot of [the men] know you underground and when they got out from underground, they didn't know you. They're the ones who talk junk about us women."

Still, most of the women miners blamed the wives for the men's attitudes, saying that the wives made it hard for the men to accept them as coal miners. One of the men said his wife "thinks it's absolutely not the place for them. She resented the fact that I worked with them, [although] she's come to where she accepts it, now that I do work with them from time to time." "The wives were so jealous," said Ellen, "that the husbands would actually go home and tell their wives that they didn't work with women." I heard several stories about the women's encounters with men and their wives at local stores, and about another incident that occurred at an elementary school.

At the grocery store, while some miners readily introduced their wives to their women co-workers, others took great pains to avoid them aboveground. Some "men were friendly when you were in the mines. They were nice and liked to help you," said Sally. "Then if you see them in a store, they would go plumb in the other aisle to avoid you because their wives were with them." As Scott also found in her 1995 study, the wives' resentment was so profound that it affected relationships among miners' children. Carla told me the following story, which still upset her.

> My son's school had coal miner's day. I brought all my stuff home, my belt and my hat, for him to wear to school dressed like a little coal miner. What's a little kid say? "Is that your dad's stuff?" He said, "No, it's my mom's." Next day he goes to school, and the little kid goes, "Your mommy's a whore." My son punches him. The principal called, and I had to go to school the next day. They called both parents in, me and her. He asks the kid, "Well, why did you do this?" That little boy goes, "Well, my mommy said she's a whore because she's a coal miner." They had a cop at the school standing by because they figured there would be trouble. I said [to her], "When I catch you on the street, your hind end is mine!" She said, "Well, my husband works with you." I said, "Your husband don't work with me!" He didn't. He was on a different shift.

She concluded, "I think they relate back to their wives' griping, raising Cain at home. Well, you're in here with him and automatically we're whores. When we enter the coal mines, we're whores."

In their communities, gossip was a common mechanism of social control meant to force coal-mining women to conform to gender-based norms. As we have seen, the normative gender system found in the colonized, working-class coal-mining communities of rural Appalachia are hypertraditional (Maggard 1994; Shifflett 1991; Scott 1995; Seitz 1995). Within the "little white man's colony," or patriarchal family, men work for pay to support family members materially, while women work at home to care for them. Ideologically, men and women's respective statuses correspond to their "essential natures." Men are assumed to be rational, competitive, assertive, and goal-oriented, while women are assumed to be emotional, cooperative, nurturing, and passive. Moreover, because gender is a master status, women are perceived and responded to primarily in terms of their "femaleness" and, moreso than men, their sexuality. By working in coal mines alongside men, women miners are viewed as violating the normative gender system within the systems of class and colony. Some members of the community label them deviant and stigmatize them according to their most salient characteristic—their sexuality. Given the complementary belief that women are unsuited—either unwilling or unable—to mine coal, they must be there to wreck happy homes.

By contrast, Michelle and Renee, the two black women I interviewed, lived in a neighboring county in a predominantly black community, where they received support from neighbors and kin. Collins sheds much light on the differences between black and white families and communities. The white model of community reflects "capitalist market economies of competitive, industrial, and monopoly capitalism" controlled by the economic interests of white males, and "stresses the rights of individuals to make decisions in their own self interest, regardless of the impact on the larger society" (Collins 1990, 52). Members of the community accept inequality and domination, either by denying them or by regarding them as inevitable. The relations of production support the patriarchal nature of the white family. But it is "unlikely that either patriarchal or matriarchal domination could take root" in black families because of African Americans' previous "relationship to the slave political economy," in which neither men nor women ruled. Black Americans continued to be exploited economically and politically after slavery, particularly in the South. This continued to influence blacks' definition of work relative to family and community, both of which were distinct from whites' "public, market-driven, exchange-based community models." Black models of community, by contrast, consist of "family with extended family, of treating community as family, and of seeing dealings with whites as elements of public discourse and dealing with Blacks as part of family business." Family income issues trumped issues of economic equality between wives and husbands, because "denying Black men a family wage meant that women continued working and that motherhood as a privatized, female 'occupation' never predominated in African-American communities" (Collins 1990, 52–53). Yet even though black women miners had more support from family and community than white women miners did, they still confronted controlling images at work similar to those that affected white women.

SOCIAL RELATIONS UNDERGROUND

Coal miners work under threatening and anxiety-provoking conditions. Historically considered to be among the most dangerous of occupations, coal mining also requires strength and endurance, regardless of whether a miner is doing heavy manual labor or operating heavy equipment. Over the past several decades, the twin forces of advancing technology and bureaucratic organization have made work and gender roles increasingly interdependent, so that the work itself strongly encourages conformity to masculine-identified norms. As a result, a "good" miner is considered to be competent and tough. A competent miner works hard and observes safe work practices. A tough miner never demonstrates fearful behavior even when he is afraid. Good miners have "team spirit," expressed through cooperation and a "give-and-take" jocularity among co-workers. Miners put great emphasis on "getting along with others" (Yount 1986), and their ritualistic behavior underground, such as teasing, making practical jokes, and horseplay, serve to reduce tension and incorporate individuals into tightly knit work groups (Althouse 1974).

Outside the mines, miners have organized themselves politically and culturally in response to coal operators' attempts to exploit them (Wardwell, Vaught, and Smith 1985). Formally, the union promotes this solidarity; UMWA "brothers" and "sisters" are united in their collective militancy as manifested in the union slogan "An injury to one is an injury to all." Coal miners are known for being "open, friendly, helping but tough; hostile to the company, but not lazy; with blunt, unvarnished feelings along with tolerance, always sharing and never cheap; everyone with a nickname, indicating individual acceptance in the group; a social solidarity recognizing individualism" (Ross 1974, 176). Older and experienced miners possess "the miner mystique—a sense of justice, toughness, manliness, respectability, pride, and, above all, solidarity" (Althouse 1974, 16). Having these qualities enhances one's reputation among co-workers and bosses, all of whom are locked into relational patterns of power and dependency.

Both male and female miners told me many times that "not everybody can be a miner." Many of them were quick to add that some men were not capable of working underground but that some women were. In the context of our conversations, being a miner meant more than just being able to do certain jobs. It also meant adopting a certain outlook on work and acting in certain ways toward co-workers. Beyond meeting the instrumental challenges of working underground, to be discussed at length in the following chapter, the women were also hard pressed to form solid working relationships with male miners and bosses. As tokens on their work crews, they posed a threat to male solidarity. As a result, at least some men reasserted their masculinity and tried to preserve their privilege, primarily by sexualizing their relationships with the women in the workplace (Enarson 1984; Swerdlow 1989). Typically, men used words and deeds meant either to exclude women or to depict them as sex objects. Reducing women to sexual beings trivialized women's attempts to demonstrate their capabilities. In the long run, the men's exaggeration of gender differences served to remind women that they had no place in the mines, that both men miners and bosses were generally united against them, and that, should the women decide to stay, men would be in charge.

Work-Related Hostility

During the first few years that most women worked underground, male miners' initial responses to them were mixed. While some men were supportive and showed them the "tricks of the trade," others made hostile remarks and were generally belligerent. "They labeled them as troublemakers right off," said one veteran miner. "All of them [were labeled because] they were taking men's jobs needed to feed their families. [The women] were just hired because the law made the companies hire them." At first, many of the men felt that the women couldn't do the work and that "we were going to have to carry them." He added that the men were also concerned because "they always said it was unlucky for a woman to be in a coal mine. It made the men feel uneasy until they got adjusted to it." In describing the atmosphere at the mine in these early years, Carolyn said sarcastically, "For some of the women, it wasn't very social." Most of the women said the men made derisive comments and complained that they simply shouldn't be there.

"They would say, 'Your husband works; what are you doing in here?' They'd give us little smart remarks and stuff," said Ellen, "because when

I was going in and a bunch of others, we was crowding in, more than one." Typically, when a sizeable number of women enter a previously male-dominated workplace, men see them as "invaders" (Kanter 1977). In the late 1970s women constituted about 10 percent of the mine's underground workers, which explains why the woman working at the diner dubbed the mine a "woman's mine." Dismayed by the men's lack of class consciousness, Carla told me, "Even some of our union brothers [have said], 'I don't think women ought to be in here. They ought to get out here and let a good man have this job.' They said we should be home cleaning house, raising kids. That it's no place for us. There were some that will even tell you they don't like to work with women." Other men tried to intimidate the women, said Sally, telling "them things like, 'There's rats in there. [They] told them stuff like that to really scare a woman from ever going in there." A few would "just avoid you. You couldn't even hardly talk to them or anything," said Lynn. "They'll avoid you and they don't make no conversation with you."

Some bosses used more explicit tactics to try and drive the women out of their jobs. "There was three of us there for a long time [in] '77, '78," said Sadie. "We took a lot of harassment. We were usually shoveling track, shoveling belt. And you had a lot of men that would want you to do all the hard dirty work while they sit on a scoop [piece of equipment]. I heard one foreman say his sister-in-law was working there. He didn't want her there and he told me, 'We tried to run you off, but he said we couldn't.'" Darlene believed that when she first began working underground she "went through eight or nine bosses, all trying to break me, make me quit." One afternoon when Rachel, a quiet and unassuming woman, and her crew were lined up for work, the boss picked her out from among the male miners to inspect the lacings on the inside of her hard hat. Measuring with his index finger, he declared that her lacings were less that the required three inches, and he sent her home for the day. Although her absence was excused, she lost a day's wages. His message, she believed, was that she lacked even the simplest technical know-how. Other women miners said that some bosses and male miners had tried to mar women's work reputations. Carla told me:

> I had put up some ventilation. I had this much ventilation left off the ground [*holds hand out to indicate approximately one foot*], and the curtain wouldn't reach the bottom. So I went off hunting another piece of curtain to attach to this curtain. [The boss]

> came up and looked, and I wasn't there. I went and got my ventilation and put it across the bottom. It was quitting time. Everybody was going to the bus, so I went and got my stuff. [The boss] didn't say nothing to me. Outside he told [the superintendent] that I didn't do my job right, [that] I'd left the ventilation like that. I went in the office and he asked me about it. I said, "Sure, he might have walked over and saw that ventilation like that, but it's not like that now." I said, "I'm on my time. I don't want nothing out of this except us three to go back in that mine and go right over and look at that curtain." We did it. I demanded we do it. They saw that it was done. I never had this problem with nobody except [the boss]. You couldn't please the man no matter what you did or how hard you worked. He just had this thing against women coal miners. Wouldn't never admit it, but it was obvious. And things got so bad between me and [the boss], they switched me off his section and put me on another section to keep down the conflict.

Lynn had a similar experience with men's collusion and lack of cooperation.

> One time I was hanging rag and the roof bolter pulled it back and knocked some of it down. One of [the men] said to the other one, "We're going to help her put the curtain back up," and the other one said, "Well, hell no." They said they ain't going to roof bolt [until the curtain is put back]. Well, I didn't give a shit. I went and sat down. So when the boss and the miner come around there, I said, "There ain't no use you pulling up in here." He said why, and I said, "Because the curtain's down." He said, "Why wasn't it up?" I said, "Well, I put it up and they knocked it down and they don't want to put it back up." So the boss went in there and put the curtain up. The boss! So the next day the shift foreman came up there and asked what happened. I told him, and he got all over them guys about that. He said, "You should have at least helped her put that curtain back up. It's everybody's job."

Black women faced a doubly hostile work environment. "I've been called nigger a couple of times," said Michelle. "Some of [the men] want to get smart with you because you are black and a woman." Renee added, "They come along once in a while, like when I first started [working at the mine]. That's the only hard time I had." The following two incidents demonstrate the double stigmatization of blacks via two powerful racial stereotypes. In the first story, black workers are associated with ineptitude and possibly sloth, as reflected in shoddy workmanship.

> We was hanging up curtain and I said, "God, you guys, that thing is sloppy," and one of the guys said something about nigger-rigging. This boss was standing right there by me and there was about ten of them right there. Immediately everybody turned around and looked at me, not in surprise at what he said but as to how I was going to react. At the time I had a hammer in my hands and I said, "Well, you know, if we was outside this mine I would let you have it with this hammer. But," I said, "you are the boss and you better say something to him." He said he caught it. I turned to this boss and said, "You're a damn liar. He didn't catch it because if he caught it then you and I and everybody standing here wouldn't have heard what he said. I'll tell you this right now: just like you are over me, somebody is over you too. You best get it straightened out. You are not going to say it. I don't want to hear no more about it." I know he said more about it because when I came to work that night, two or three bosses came up to me and said, "I heard what was said to you," [but] I spoke no more about it because you calling it a nigger rig don't bother me because I am not a nigger.

In the second account, blacks are assumed to have superhuman or animalistic strength, which also conjures up a "slave" image.

> This one guy said, "Them niggers sure can lift them blocks." He was sitting down and I was standing up. I just took one of my feet and kicked his chair. No one said a word. He said he was just repeating what someone else had said. The boss was sitting right there and didn't say a word. So a few days later one of the guys said, "I saw the expression on your face. You was mad enough to fry an egg on top of your head." So the boss was standing there, and I said, "You should have said something." He said, "For your information, I have had a talk with these

> guys about that." The ones who was with me said, "Oh no you haven't neither, because I haven't heard it." So then at lunchtime he brought us all together and the boss said, "I need to bring this up. Something was said and somebody got upset." And I said, "That's right. I got upset because he said 'nigger.' That's right, and I'm not going to stand for it." This boss wanted to say I got upset over nothing. So he said, "I've had this talk about racial names and I'm not going to put up with it." But he should have said it then.

Michelle allowed that the company "did not put up with it." But even so, she said, "I deal with my situations right there on the spot. I tell them, anytime you want to jump, white boy, whatever you're going to do, finish it right here." One time, she recalled, "this one man said, 'I'm a man and I don't hit a woman, so why don't you bring some big black guy to fight for you?' And I said, 'Why should I get some black man to fight for me when I can find some three hundred white men who can't stand your butt?' He's a bully. Like I say, it's very few."

In meeting the challenges they faced in the mine, several women justified their presence by emphasizing their practical needs.[5] Ellen,

5. As reviewed by Seitz (1995), Moser (1989) distinguishes between practical and strategic needs and discusses their relationship. Within the existing gendered division of labor, women's practical needs include providing food, clothing, and shelter for themselves and their families. Practical needs "are formulated from concrete conditions women experience, in their engendered position within the sexual division of labor, and deriving out of this their practical gender interests for human survival." Their strategic needs are political in nature and involve challenges to age-old male-dominated institutions and their supporting ideologies. They are "formulated from the analysis of women's subordination to men, and deriving out of this the strategic gender interest identified for an alternative, more equal and satisfactory organization of society than that which exists at present, in terms of both the structure and nature of relationships between men and women" (Moser 1989, 1803). Often, women have had to attend to their strategic issues in order to meet their practical needs (Seitz 1995).

Just as coal miners' wives have become active in strikes (Giesen 1995; Maggard 1990), other working-class Appalachian women went to work underground, despite their fears and the cultural taboo, to provide for themselves and the welfare of their families. Often they have faced "the dilemma of balancing a response to economic transition and hardship with deep-seated local affiliations and loyalties" (Tickamyer and Henderson 2003, 112). Feminists working in the area of planning development assert that because of their marginalization and collective grassroots resistance, working-class Appalachian women actually have more in common with many women in Third World countries than they do with their middle-class counterparts in the United States (Seitz 1995). For example, upon entering the paid labor force, both have become unwilling participants in the manufacturing of gendered hierarchies (Anglin 2002; Ong 1987; Tallichet 1995). Like poor women in other countries, working-class Appalachian women have had to balance their practical and strategic needs in order to make ends meet at home.

newly divorced, had once quipped that, since "it was a man that put me here," she felt she had every right to a man's job. When Sadie first started working underground in the mid-1970s, one of the male miners asked her, "Why don't you go home and give this job to a man that needs it?" She replied, "When I come up this hollow to get my job, they was begging for men to work. Well, they didn't come and get it. It's mine, and I'm keeping it." In both cases, a man's failure gave women the right to "work like a man." Lynn, also newly divorced with children, told me, "I even had a boss tell me at one time he didn't like to work with women, and he wanted to know why my dad let me come in the mines. I said, 'Buddy, I was twenty-eight, divorced, and single. I could do whatever I wanted whenever I come in.' And he said, 'I just don't like to work with women.' And I said, 'Well, you just best get your dinner

---

Seitz (1995) also points out that grassroots mobilization among women marginalized in different areas of the United States and beyond, including rural women in central Appalachia, posed a challenge to the patriarchal elements of capitalism. These movements have been regarded as engaging in "radical practice." Although they have operated outside the state's institutional apparatus, they have enlisted its assistance in pursing women's strategic interests. In addition to infusing democratic practices into the everyday operations of corporate capital, they also serve women's practical needs by providing economic opportunities for their own self-reliance. Social mobilization such as this relies on women's continued resistance at the local level, which can have lasting results on the region's sexual division of labor and the traditional gender ideology that reinforces it. Those women who obtained mining jobs and kept them represent just such a social force.

In resolving their strategic issues in order to meet their practical needs, Appalachian women miners enlisted the state as an ally. Although the state often acts to support the material interests of capitalists and the ideological interests of patriarchy, it also serves to mediate conflicts as they occur between them or arise from them (Eisenstein 1984). Until recently women were subordinated in terms of their political representation. But as more and more women have entered the labor market, they have become more aware of how institutions constrain their roles in the labor force and keep them dependent on men. Increasingly, women have demanded higher wage-earning and decision-making status as workers. Relatively speaking, this poses a great challenge to patriarchy, and to a lesser degree to capitalism, while also drawing them into conflict with male co-workers. Women's groups have been instrumental in gaining the support of the state on behalf of women coal miners. The state has responded to their pressure with the enactment and initial enforcement of federal antidiscrimination legislation, threatening coal companies with the loss of federal contracts, profits, and later with lawsuits over sexual harassment and sex discrimination. Again, the interests of the capitalists (as defined by the threatening actions of the state) were brought into direct conflict with the system of patriarchy (Sokoloff 1988). As a result of the state's pressure, women were able to get and keep coal-mining jobs in record numbers, despite male miners' protests that women were taking "men's jobs" and that men would have to step in to do the women miners' work for them, thus driving up the cost of coal. Such beliefs were an attempt by men miners to realign capitalist interests with their own patriarchal ideologies about male dominance and superiority.

bucket and go the house, because I'm here to stay, and I'll be here when you're gone.'"

But Lynn also said that when she could, she simply avoided some of these men. "There have been a few of them that's said, 'Well, I really don't think women's got no place in the mines, but they're here,' or something like that. But they're not being smart about it, they just tell you their feelings, and when they do, I just kind of stay away from them. I think, well, that's their right. But my right is here to work and I'll just do my work and not bother around them." Rather than defy the men directly, she sent a silent message that she didn't need their help. This strategy could backfire, however, if a woman became socially isolated and was hampered in becoming socialized to the job and learning new skills.

Coal-mining women challenged men's breadwinner status within the patriarchal working-class family. But by presenting themselves as "reluctant breadwinners," they were often able to neutralize that challenge. This was difficult for women who had no children, however. When Kelly first started working underground, a male co-worker asked her why she wanted to be a miner when she had only herself to support. Much to her surprise, he asked her if she was a lesbian. She replied, "What difference does it make what I tell you? You already have your mind made up." She reasoned that men were not only intimidated by a woman who could handle a coal-mining job but also by the possibility of their homosexuality. As Ellen put it, "Those men up there probably think we're all lesbians!" A woman who could be both financially and sexually independent of men threatened the "natural order" by undermining heterosexual men's power and privilege (Schur 1984). In sum, women miners were still regarded in sexual terms, mining was still seen as "men's work," and women still did not belong.

The Sexualization of Work Relations

Work relations were "sexualized" by behavior that expressed the "salience of sexual meanings in the presumably asexual domain of work" (Enarson 1984, 88). At the very least this behavior took the form of male miners' language and jokes, which most women accepted and even occasionally participated in themselves. Having grown up with coal-mining fathers, uncles, and brothers, most women felt that this kind of behavior typified the masculine-identified culture of coal mining. Most said that jocularity was one of the best "ways to get through your day," and they saw sexist jokes as part of their socialization to mining, as I discuss more fully in Chapter 4. In conditions of relative oppression, where the development of a dual consciousness is necessary, it is not uncommon for members of the subordinate group to adopt the thinking and manners of the dominant group (Collins 1990). Moreover, sexual harassment is in the eye of the beholder; it is harassment when the recipient sees it as offensive, objectionable, and unwanted. It is less about sexual desire than it is about male domination over women (MacKinnon 1979; Reskin and Padavic 1988; Uggen and Blackstone 2004). As with the men's work-related hostility, most women experienced sexual harassment, which made the work environment that much more hostile. The women I interviewed estimated that during their first few years on the job, at least half the men miners and bosses engaged in verbal innuendo, sexual body language, social derogation, sexual propositioning, and sexual bribery. (For a discussion of different types of sexual harassment, see Gruber and Bjorn 1982.)

The women were explicitly reminded that they were intruding into a previously all-male world rife with the sort of ritualistic buffoonery and vulgarity meant to reinforce male bonding and solidarity (Vaught and Smith 1980). The following stories exemplify versions of male bonding that simultaneously "otherize" and "sexualize" women so as to reaffirm men's dominance. For example, Ellen said:

> My God, I worked with a [foreman], and if you want to work with a bunch of men, that was nasty. They was pretending they was queers in front of me. You wouldn't believe it, it was like one was humping the other one, but they had their clothes on. This is what the boss said: "You scared of us, ain't you?" I was so mad, it hurt because of the sickness of it. I said, "No, I'm not scared of you all." And he said, "Well, Ellen, this is our little world down here and you don't belong here." I said, "I thought [the company] owned this. Ain't all of us in but one company?"

On another occasion, she told me:

> Most of the time I have never run up on a man using the bathroom. They're like the women, but this one guy that was always pestering me, he didn't want me there, I don't think. He was

> zipping his coveralls up and it was like where everybody walked past, all the men, but I was the only woman up there. And he said, "You almost caught me." And I said, "Uh huh." [Then he said,] "You might've liked what you seen." I said, "I doubt it," and walked on. But it's stuff like that really burns you up.

Other women also remarked that some men often urinated in front of them or grabbed their genitals, pretending to be caught off guard. Sally said:

> Some of them actually got their things out and played with it. I would just avoid it and go on. I would go on about my business. I didn't tell anybody or anything. They was there. Let them do what they wanted to and let them look stupid. As long as they didn't touch me, I didn't care what they done.

Carolyn told me how a male co-worker made a more casual, if not more subtle, habit of exposing himself:

> Zipper down. He was wearing a pair of jeans that was split all the way up. I didn't even notice it until this friend of mine, she worked with him, said to him, "You know, you need to wear something else to work." Then he starts working with me. I was pretty cool about it. When any men would come around, he would button up his britches.

Later, the same man was reassigned to a job that required wearing a rain suit, so she brought him a pair of pants. "I went back down there and said, 'Oh, by the way, your pants are ripped. You need to put these on.' He just looked at me and then put them on. That's the way to handle things with men," she said. But not all the women chose to handle men's sexualized behavior with the same cool detachment. In another incident, Sally attempted to undermine her male co-worker's masculine prowess in order to shut him up.

> They would tell you how many inches they had and [say,] "I bet I can satisfy you. I bet I can do for you what your husband can't do." And they had a nice wife sitting at home. Well, I got back at one of them. This one stayed on me all the time and I said,

> "Buddy, do you know what that thing looks like? Have you ever seen what the turkey's neck looks like when they pull it out at Thanksgiving?" I said, "That's what your thing looks like, and when you're riding your old lady, do you know how many pieces of fly shit she counts while you think you're being a macho man?" He never did bother me no more, but that's what it took to shut him up and [get him to] leave me alone.

A similar incident showing the solidarity between a male miner and a boss involved a sexual prank at a woman's expense. During her first year at the mine, Jessie, a self-assured athletic woman then in her early twenties, had been assigned to a job that required her to bend over much of the time. Unbeknownst to her, every time she did so, the boss and another male miner would shine their cap lights on her "hind end." Eventually another man on her crew told her about it. She confronted them both, requested a transfer, and walked off the section. Fearing that she would "take him to the office," the boss followed her out of the section, apologizing profusely and promising to give her a good recommendation. Explaining her anger, she said, "I'll tell you, I don't like to be treated like a plaything. That's what made me so mad." Conspiring man to man, the boss granted the miner permission to sexualize Jessie, negating the hierarchal differences between them and reasserting their common gendered interests. But, aware that her objectification also trivialized her as a coal miner, Jessie was determined to stop it. Interestingly, another male co-worker had refused to play along, suggesting his belief that solidarity among miners was more important than the male bonds of patriarchy.

Social derogation occurs when "a woman's reputation [is] demeaned by men who spread malicious rumors or stories about her" (Gruber and Bjorn 1982). The mine was often a regular rumor mill, and stories spread like wildfire among men and women alike. Social derogation only exacerbated the effects of a woman's token status by increasing her potential for becoming socially isolated. For example, Carla's experiences with a boss who was eventually fired for unsafe working practices pitted her against her male co-workers.

> One time [the boss] went and told the guys behind my back that I had sucked his dick, is the way he put it, down the jack line. It came back to me about a week or so later, and I got mad

> about it. I didn't say nothing until everybody got to the bus. I walked up to him and I said, "[Boss], I want to ask you something." I said, "When did I suck your goddamned dick down the jack line?" He goes, "I don't know what you're talking about." I said, "You're a goddamned liar." I said, "You told every one of them and you didn't think that they'd find out I'm not doing the shit you said I was doing, and come back and tell me things, did you?" [He said,] "Who told you this, I'll take them to the office." And I wouldn't tell him. Right there it proved to the guys [that the boss was lying], because some of them actually believed it!

Because gossip is an informal mechanism of social control, it defies formal regulation. Informally, it held women's behavior in check. "If you've talked to somebody, they automatically got you sleeping with them," Carla said with exasperation. "[The crew] has me sleeping with that man, and it gets out all around the mine." Such gossip solidified the controlling image of the woman miner as a whore, an image that all women miners were then obligated to resist. One way of dealing with gossip was to challenge the source of the rumor directly, as Carla did. But this was not always feasible. The men's word was considered more legitimate than women's. As Carla recalled:

> I went through pure misery for about a year because the boss hassled me constantly, he lied on me constantly. I mean, he lied to the crew that I worked with, telling them I was going to take them in the office and I was going to say this and I was going to say that, and I didn't even know what was going on! I knew they all buttered up to the boss. He was saying he could get the guys to say I did [sexual acts]. You be nice to them, give them easy jobs, you butter them up. They're going to say anything you want them to say.

Social derogation could also affect a woman's working relationship with her crew, particularly when it involved a boss. Not only was a crew's work highly interdependent, but a high degree of dependency exists between the boss and his crew. On the one hand, miners depend on the boss for their breaks, work assignments, and rewards, such as promotions. Work assignments and promotions stem from the boss's estimations of them as miners. At the same time, the boss depends on his crew to get the work done, and their performance affects his reputation with his superiors. What miners and bosses believe about each other thus profoundly affects a miner's work reputation and job status. In Carla's case, rumors had dire consequences because the foreman's deception resulted in her crew's lack of cooperation.

> I didn't even know why everybody all of a sudden quit speaking to me, giving me the cold shoulder. The plow would be coming to the head when I worked for [the boss]. Coal, if it gets built up enough, it's going to gob off over into the line. When the plow got to my set of jacks, I would get on the phone and holler at the head operator, who is a union brother, to cut and hold the plow until the coal cleared up. It wasn't done because he had this guy pissed off at me. Here I worked myself to death keeping it shoveled up because he felt that way.

Malicious gossip only made Carla's job more difficult as she became more socially isolated from her crew members. In this way it ultimately reinforced the perception that women were not cut out to be coal miners.

When I asked Darlene how the women were treated when they first started working at the mine, she snorted, "Like a piece of pussy." Ellen said that one of her male relatives, who also worked at the mine, told her "that a boss said that all the women made beds out of rock dust for the men. You know, like that's all we did was go in there to sleep with them?" With a hearty guffaw, she continued, "I told him, 'Well, you tell him that we got all the beds made and we ain't got no room to put him up there, so there's no sense him coming in there.'" Once again, the controlling image of whore is quite evident. It also corresponds to the larger culture's "women-as-prostitutes" paradigm. In a society where women are subjugated, their survival depends on their construction of self as a sexual being, including being passive sex objects or providing sexual favors in return for social or economic rewards (Silverman 1976). These expectations are heightened within the context of the patriarchal family and further supported by the working-class norms and values of a colonized region such as central Appalachia. I, too, witnessed these sexual expectations during my stay in the community. One evening, while talking with the women who worked at the diner, a male patron began complaining about the increasing cost of supporting his family. He said he wished he were a woman, because then he could earn an easier living as a prostitute. A younger woman working the counter quipped, "Don't quit your day job." Amid our laughter, he reasoned aloud that she was right. After all, if he wanted to have sex, he would simply go home to his wife, because "it was cheaper to keep her." In other words, alimony was costly compared with his wife's unpaid labor, including her sexual services.

Sexual propositioning and sexual bribery were also common at the mine. Sexual propositioning posed no particular problem for some women because, as Jolene said, "A man is too proud to keep asking, if you're not going to go. If you made it known, they're not going to come back and keep begging you to go out." She added emphatically, "If I ever thought somebody was discriminating against me because I had turned them down, I would raise all sorts of hell." But both women and men knew that a boss's sexual propositioning usually amounted to sexual bribery, the "quid pro quo" form of sexual harassment. Until the company rule was established, some bosses made women's work assignments contingent upon sexual favors. An older male miner told me, "If the guys asked her out and she wouldn't go out, why, they wouldn't help her as much as they normally would. If the foreman asked her out and she wouldn't go, the foreman would make sure she got a dirty work assignment." Sally remarked:

> [Some bosses] actually gave [women] the pitch that they sleep on the couch [and] that they're not sleeping with the wife. You know, we're not getting along. Hey, I'm a woman. They're not going to tell *me* that junk. I mean these bosses really pull the wool over these women's eyes. They don't just pull the wool, they take the blamed sheep and pull it down over their face. These women are actually stupid, but they do get the easy jobs. So I had to work like a brute because there wasn't nothing there I was going to have no affair with.

She concluded, "It was like a game with them to see how many of those women they could get into bed." Three days after Ellen had started working underground, her foreman took her to a dark and isolated part of the mine. "He was trying me out," she said, "because he kept [turning his cap light off]. He seen he wasn't going to get no place, so he took me to a boom, and he left me there to shovel" with another man. "And this guy said to me, 'I'm going to tell you something. If you let these bosses pinch your titties, you'll get along. But if you don't, you're going to get the awfulest job here in the mines that ever was.'" Carla, who had once refused her boss's sexual advances, said, "He just made things terrible for me. He came right out and told me, 'If you don't sleep with me, you've had it, girl.' He worked me like a dog. I got every shitty detail there was." In avoiding a foreman's sexual advances, Carolyn admitted that she hadn't "bid on jobs before because I didn't want to cause problems. This man was married and I wouldn't go out with him. It was to avoid a bad situation. See, the boss was pushing me."

Other bosses took a more indirect approach in order to conceal their intentions and avoid being punished for breaking a company rule. When Jessie was a red cap, or trainee, her boss, who wanted to go out with her, asked another woman to approach her for him. Soon after this friend conveyed her refusal, Jessie was reassigned to a more difficult job, which she believed was punishment. Another situation involved Irene and her daughter, who were both working at the mine. Her daughter's boss wanted to go out with Irene's newly divorced daughter. But the boss "was married and had a baby! You don't fool around with that stuff," she said. "Well, she wouldn't go out with him. He wanted to put pressure on me to get her to go out with him." Irene complained about it to an older male miner, a union official at the time, who explained,

> Her daughter was a divorcee, and she went in the mines up there. She was an attractive young woman, and the guys would hit on her and hit on her and hit on her. Well, there was this foreman that really wanted to go out with her, and she didn't want to have anything to do with him because she was going steady at the time. He kept harassing her at home and on the job. And one night he even put [Irene] in a water hole deliberately because her daughter wouldn't go out with him.

After he told the mine superintendent, "the next day that foreman was gone. You can't put up with that. [The superintendent] wouldn't tolerate it at all." The same miner said that for the first few years after the women began working at the mine, several foremen "hit on most of them, even the married women. We had a foreman get up and testify that he paid women wages for not even showing up. He'd pay them a full day's pay. As quick as he got to the mines and things would settle down, he'd leave and go visit them at their homes. [Later, he would] come back to the mines." But over the years that followed, these situations finally "blowed out," he said. "The ones that were involved in it are gone. They no longer work up there." Nevertheless, knowing that they could get easier work in exchange for sex caused suspicion among the women over job assignments. Unless she had a solid reputation for rejecting sexual propositions, a woman who got an easier job or a promotion became an object of suspicion. This kind of doubt also eroded the women's solidarity, as we shall see in Chapter 4.

Several women said that when they began working underground they were less apprehensive about the mine's dangers than they were about the men's behavior. "Maybe the men had never worked with women before, and the only job they ever had was in the coal mines," Carolyn said. "When I went in," said Ellen, "I was a little bit afraid of them. A lot of the men would ask the women to go out. I can't tell you how scared I was. I wasn't scared of the mine, I was scared of the men. See, I don't work with women, I work basically with all men. I'm not scared of them now, but back in them days, I was scared to death of them. When they got you down there, they can do anything they want to."

Many of the women devised a strategy for dealing with men from the start. They purposely projected a brave persona and emphasized that they'd taken a mining job so as to meet their practical needs. "When I first came here I set myself up right away. I've made it known: 'Don't bother me. I'm here to work, I'm not here for romance [but for] finance,'" said Kelly. "'Don't get in my path, I won't get in yours.' Once you establish yourself, they know your boundaries." Michelle stated adamantly that a woman working underground had "to be strong when you first come in. If you're scared, don't let them know you're scared. If you're going around thinking about how they're going to treat me, what's going to happen? Don't let them know that!" She continued, "Even if they intimidate you, don't show them a weak face. I have actually seen two women, if one of the guys say something to them, they are crying. I see it as a sign of weakness." She said of herself, "I am a strong woman. I think every other woman should be strong, and I cannot stand to see a woman whimper." Renee added, "You got to come in there like I know what I'm doing and if I don't, I don't want them to know. Strong from day one. I don't care if you go home and cry, but don't cry while you're in there." These women consciously projected an image of emotional strength that deliberately thwarted the men's stereotypical expectations. This demonstrated their dual consciousness under oppressive conditions as a means of everyday resistance.[6]

Notably, most of the women's solutions to these daily social problems put the onus of responsibility almost entirely on themselves. Most of the women I interviewed declared that if a woman was harassed, no doubt she had asked for it. "The majority of the men up there are good to you if you let them," said Lynn. "But they'll treat you how they see you act. You go in there acting a floozy, they'll treat you like one. You act decent, they'll treat you decent." Carolyn expressed the same faith in projecting a proper image and in the norm of reciprocity.

6. Several socialist feminists have noted that oppressed groups possess this duality and that it constitutes a complete and privileged knowledge about their world and social relationships (Collins 1990; Harding 1987; Hartsock 1983; Jaggar 1988; Nielsen 1990; Smith 1987). Collins (1990) contends that the development and use of a "dual consciousness" indicates their everyday individual and sometimes collective resistance against class exploitation, gendered subordination, and racial oppression. Being compliant while simultaneously harboring an understanding of one's oppression can provide the basis for an "everyday form of resistance" (Scott 1985). The recent work of several feminist scholars supports the idea that quiescent behavior itself can be a form of resistance for black, rural, and Appalachian women (Collins 1990; Sachs 1996; Seitz 1995). They maintain that if we regard women's resistance only in the traditional sense, in terms of collective and institutional responses they have made, we will miss an opportunity to understand it in its fullest individual and personal sense. As Sachs (1996, 26) explains, "one way to reconfigure women's potential is to note that resistance is more than organized struggle. The daily activities of their lives shape many women's opportunities for resistance. Women's acts of resistance against oppression take many forms, often unseen or underappreciated for their cumulative effects."

Seitz (1995) makes an excellent point regarding the utility of the Marxist concept of false consciousness and its alternative, class consciousness, relative to the term "dual consciousness." According to Marxist theory, workers, regardless of gender or race, are subject to capitalists' exploitation and their means for collective resistance lie in the extent to which they cast off their "false consciousness" about the capitalist system and their role in it, and act on the basis of a shared "class consciousness" to rebel against controlling elites. Marx believed that all forms of oppression began with class oppression. Recent feminist scholarship on the subject rejects this idea as one-dimensional. For example, Seitz (1995) asserts that it represents a dichotomy, an either-or situation, and fails to consider the complexity of the consciousness of oppressed group members. Marx did not take into account the fact that members of oppressed groups are simultaneously aware of who they are on their own terms and the expectations elites have of them. Moreover, according to Collins's (1990) matrix of domination, individuals confront multiple forms of domination simultaneously. As several researchers have acknowledged, gender and race are also central to any cogent analysis of relations within the contexts of work, unions, and the state (Fonow 2003; Maggard 1999; Scott 1995; Seitz 1995). Indeed, Maggard (1999) found that sexism thwarted the emergence of class consciousness, contributing to the failure of a hospital workers' strike in the heart of Appalachia. That gender (or race, or both) can fracture class consciousness underscores the unique nature of the former at the expense of the latter.

> I think the men just wanted to see just how far they could go with the women. I think the men thought they were in there for romance. So you got to set them straight. If you were a respectable person, they treated you with respect, because it all depends on how you behaved yourself. It doesn't matter where you work, it's how you conduct yourself, wherever you are. You can get along and kid with men and they get along with you and they get respect from you and that's just how they do it.

When a woman enters a male-dominated occupation such as coal mining, she is often, if not always, regarded as a representative of her sex. Thus, the sexual behavior of some women reflects upon them all, and most women feel a social and moral responsibility to the whole group. Because of "some of the things that have happened there," said Carla, "they have ruined the name for a woman coal miner." During my discussion with Kelly and Alice, who had worked together for years, Kelly told me, "When Carla went to the doctor's office, two women said all the women [working at the mine] are sluts. That's not so! I'm not." And, pointing to Alice, she said, "She's not. We're categorized." Ellen said, "Sometimes there's been a few cases like the boss and the worker, you know? We got one woman, she's married to a boss. She broke his home to get him. That makes us all look bad, and the [other] women don't like her." Not only did one woman's infidelity tarnish the image of the others, it also affected other women's work experiences with certain men. Carla said, "[My boss] wanted to sleep with me. I wouldn't have anything to do with him. He thought if a woman worked for him she had to sleep with him, because there was one woman working on the section already [who] was sleeping with him. It was obvious. Everybody knew it. They didn't hide it. And when it came my turn, I wouldn't sleep with him." Her refusal resulted in a more difficult work assignment.

Culturally, gender is dichotomized, constructed dialectically, and "doing gender" is unavoidable (Parker 1992; West and Zimmerman 1987). As conditioned by colonization and class, the Appalachian subculture of coal mining is more traditionally patriarchal than in other regions of the United States. The subcultural expectations for women's and men's behavior are more exaggerated, and members of each sex are held more strictly accountable for their gendered performances. Only two decades ago, Anglin (1983) found that Appalachian women who allow themselves to be left alone with men are seen as granting men permission to be sexual with them. Women who do not voluntarily segregate themselves from men deserve whatever consequences befall them. This double standard for sexual behavior is grounded in patriarchal ideology about the differences between men's and women's "essential natures." When Lynn was discussing the relations between men and women underground, she remarked, "See, men, they tend to watch women more. I believe it's just the male in them. They'll watch you and see how you act." As with some of the other women, her acceptance of patriarchal ideology became the basis for her resistance, which allowed her to negotiate her identity as worker and as woman. But the double standard also led to "blaming the victim" of sexual harassment. Speaking of sexual affairs at the mine, one veteran male miner told me, "I think a lot of them egged it on, asked for it. I've been in groups where I saw this happen, where it seems like that's what they wanted the men to do. In general, I don't know that any of them was abused in any way . . . because at any time, if they would've came to me or management, if they want to get it stopped, they can get it stopped." Or, as the superintendent said flatly, "They could invite it and then cut it off."

Yet another reason the women assumed much of the responsibility for their relationships with men involved the formal norms governing workplace relations. Before the women arrived in large numbers, the company had few rules governing miners' behavior. By the early 1980s, however, the company had issued rules strictly forbidding any kind of harassment, horseplay, or profane or obscene language. Jolene said that management officials at the mine "told the foremen, 'If anything comes up we will take the women's word for it.' They were told, 'You watch what you do. Don't get yourself in that position.'" Now, she said, "they're not going to bother anybody. I mean they'll ask you out, you say no, and that's it. I think it probably cooled the foremen off a lot." Although the company had now provided the women with an official channel for stopping any form of harassment, the way the policy was implemented put the women in a double bind because of its "damned if you do, damned if you don't" consequences.

Specifically, it was the women themselves, and not other men, such as foremen, who were solely responsible for reporting harassment. Some women were reluctant to report harassment because this created tension among crew members. A few others said that when they reported harassment, they were transferred to another work location, while the man who violated company policy stayed behind with the crew. Such action reinforced women's stigmatization by defining the woman as the problem and, in effect, rewarding men for harassment. In this way male privilege and status were preserved. Moreover, reporting harassment to management violated the UMWA oath of solidarity taken by all union miners and this defeated the women's efforts to integrate and fragmented their class solidarity. Women who reported harassment also served the interests of capital by giving bosses an opportunity to head off any grievances or lawsuits women might file. In the interest of class solidarity, Carla staunchly declared that she "wouldn't say anything against my union brothers because that's not right. You're not supposed to do that. We took an oath, and I have stood by that oath. I have not said anything to any of them bosses about anybody up there." The company's policy was divisive, she said, because it was another way that the company aligned itself with male miners against the women. When a woman did report harassment to management, miners referred to it as "packing the mail" or "carrying the mail." "You just don't do that," said one of the men, "but management did put the women in that perspective. They said, 'Hey, you got any problems with people bothering you, you come to us.' They didn't think of the union, they said come to us." Fortunately, most women knew they could go to the union to stop harassment. In the case of Irene and her daughter, discussed above, Irene went to her union "brother" and the superintendent swiftly dismissed the offending boss. Like Carla, her sense of class-based solidarity was very strong, and she refused to allow the gendered nature of the situation to fracture her class solidarity.

The company's policy also stereotyped women by assigning them "protected" status, thus adding to their stigmatization as "other." The superintendent complained to me that many of the women hired during the late 1970s made "ridiculous demands" on the company over social and sexual matters underground. As a result, he said, the bosses became distracted with the women's problems because they felt obligated to protect the women miners. "If the foreman was weak," he explained, "he could have half the crew bird-dogging" a woman on his crew. Interestingly, some women denied needing the company's assistance, recognizing that management's intervention could make matters worse. Handling matters themselves, as Michelle indicated, gave them a sense of empowerment and challenged the men's power. The superintendent also remarked that because the company had to take the women's concerns seriously, management had to "teach the men what harassment was." Previous research about women's integration into the workplace has shown that "men are so accustomed to viewing women as sex objects that they often find it difficult to accept them as coworkers" (Gruber and Bjorn 1982, 273). Following my own underground tour of the mine, I remarked to the superintendent that the harassment rule was also important for safety reasons. Turning to the mining engineer who had given me the tour, he said, "We're still protecting them, aren't we?"

Overall, the women's experiences with and responses to men's work-related hostility and sexual harassment varied. No one type of woman was targeted for this treatment. Rather, the men's ill will and corresponding treatment depended on the men's attitudes and how women reacted. Although a few of the women said they had never experienced sexual harassment of any kind, they allowed that they would readily report it if it occurred. Some maintained that sexual harassment was more media hype than anything else. Others told me they had reported it to either a management or a union official or had resolved it themselves. Some women responded successfully with vulgarity and off-color jokes of their own. Finally, a few who tried to reason with men, or who responded with anger and tears, found that they were simply targets of further harassment. But as the years passed, virtually all the women said that the men's hostility and sexual harassment no longer concerned them. They expressed their faith in the company's policies and insisted that harassment was infrequent. The company rule "helped quite a bit . . . [because] they're sort of afraid," said Kelly. "I think things have changed so much since the first woman come into the mines. She was harassed *a lot*. But then things have changed because they've accepted us." Carla agreed that sexual harassment was for the most part over with, but mostly "because after all this time, they see they're not going to get in my pants, and they just sort of leave me alone. Eventually I guess they just get over it, they start talking to you, but they don't harass you no more for sex." As the men became more accustomed to women underground, the more overt forms of sexual harassment "settled down," said one male miner, because the women "really aren't new toys anymore." Similarly, some women said that although sexual harassment had diminished, it hadn't disappeared completely. As Jessie concluded, "I think it's still going on, but it's just more subtle now."

Like other researchers (see Reskin and Padavic 1988), I found that when even a few men harassed or degraded women, this affected how the women felt about their jobs. Despite these issues, women who have well-paying jobs are more likely to stay in those jobs than women in lower-paying jobs (Reskin and Padavic 1988). For working-class women in central Appalachia, there were virtually no alternatives to coal-mining jobs that paid as well. Without a doubt, some of the women have suffered the emotional consequences of harassment. Carla recalled, "I was aggravated. I'd come home, I'd cry. I couldn't sleep, my nerves were tore up." Nonetheless, the women miners I interviewed said that the men's hostility and sexual harassment actually strengthened their resolve to stay. As Carla put it, "They wasn't going to push me away from my job!" Lynn, who was "bull-headed" when dealing with difficult men, repeatedly stated that "the women have to stand up for their rights." As other researchers have shown, moreover, sexual harassment does not necessarily have a negative effect on advancement (Gruber and Bjorn 1982; McIlwee 1981; Walshok 1981). I asked Carla, who had advanced to an operative job, whether harassment dampened women's aspirations to advancement. She shot back, "If we have the seniority, and we bid, they have to give it to us." In work settings where advancement is formally regulated, sexual harassment may be less consequential for women determined to move up than it is in unregulated workplaces.

In the coalfield communities of central Appalachia, the working-class patriarchal family has served the interests of colonizing capitalist forces for generations. Within these families, men have been the traditional breadwinners. Having confined their view of women to the roles of mother, aunt, sister, daughter, wife, and mistress, the men were understandably unsettled by the advent of women working side by side with them underground. Because women were now doing "a man's job," they challenged the presumably natural order of gender relations, and so were held accountable for breaking through this age-old boundary (West and Zimmerman 1987). As Swerdlow (1989, 381) has noted, "men have a status stake in the sexualization of the workplace when the division of labor renders women equal to men" or at least has the potential to do so. Thus, men's sexual harassment of women workers was less about sex than it was about maintaining male dominance (MacKinnon 1979; Reskin and Padavic 1988; Uggen and Blackstone 2004) and reaffirming the subordinate status of women. Objectification and work-related trivialization are mutually reinforcing processes (Schur 1984, 142). As Jessie succinctly put it, "The men look at our bodies and not at what we can do." Unfortunately, these processes also lead to the gender-typing of jobs and a gendered division of labor when women are channeled into "women's jobs" underground, a subject to which we now turn.

2

FROM RED CAP TO COAL MINER: ADAPTATION AND ADVANCEMENT UNDERGROUND

During their early years, the women's adaptation to work underground was as much physical and psychological as it was social. As they adjusted to the physical demands of their jobs, they also learned to cope with the dangers on a daily basis. Coal mines are noisy, dirty, and damp, and their hazards are unpredictable. Problems with access, ventilation, lighting, drainage, roof supports, and coal extraction and conveyance are never ending. Accidents nearly always involve earth, fire, water, methane gas, or some combination. Roof falls are the leading cause of death, and miners warn one another about boulders or "widow makers" that can fall from unsupported top overhead. Sally explained that even if the roof is secured by bolting the overhead layers of rock together with roof bolts or "pins,"

> that top will fall, and I have seen the pins actually pop off and hit the ground. [Or] the pin will stay in there but the plate [and] the top would come down. They hit the ground and there's nothing up there but that pin. There's no plate. It's on the ground. They tell us to stand between two pins? Not this old girl, she'll find her a nice place to stand. It wasn't even safe to stand under a pin because the top would give and the plates would fall off. Really, none of it was safe when you come right down to it.

Explosions are also a persistent threat when levels of methane gas or coal dust build up and get ignited. Even if miners survive the initial blast, they can still suffocate as the oxygen in the mine dwindles and is replaced with deadly gasses. Certain areas of gassy mines are plagued with deadly "blackdamp"—pockets of oxygen-deficient air. Miners told me that their mine was "very gassy." "You can get sick standing right in the face just about any time with all the air we've got," Lynn said. Injuries from operating high-powered equipment using high-voltage electricity in tight working areas include twists and joint sprains, broken bones, dismemberment, and hearing loss. Finally, after working in coal mines for years, miners face the possibility of contracting coal miner's pneumoconiosis, commonly known as "black lung."

Miners have described the mines as "eerie, weird, or spooky" (Yount 1986,. 173). Without the light from miner's headlamps, a coal mine is completely dark. New miners can easily become disoriented and wander off in the wrong direction. Several times I heard a story about a miner who lost his way when his cap light went out. Hours later, a search team found him in an isolated section of the mine some five to six miles from where he began. The temperature underground is usually stable, unless the air vented into the mine makes it hot or cold depending on the time of year. The noise from machinery makes conversation difficult. Otherwise, miners work in silence punctuated by the sporadic loud popping and cracking sounds coming from the settling of the rib or the low rumbling sound of the roof "working," meaning that the rock slabs overhead are shifting. Ten minutes into my own underground tour, I heard the unmistakable thunderous noise of the roof settling overhead. Carolyn and I agreed that it sounded like a cloudburst. Timbers that are used to support the roof may occasionally creak when taking on additional weight from above. At times, miners have also reported feeling the area of the mine around them shake or bounce.

Since the 1960s, the mine's operations have expanded, and problems with methane gas prompted the construction of more ventilation shafts. Roof falls and popping ribs have also caused their share of injuries over the years. In response to concerns over mine safety, the company began having mandatory safety meetings every few weeks to discuss the dangers associated with ongoing operations. Knowing something about these hazards gives miners a sense of control and reduces the tension they feel about working underground (Althouse 1974). While most miners refrained from showing fear, they admitted to being apprehensive about mine work. The apparent paradox allowed them to cope with constant threats of serious injury or death. Their demonstration of outward calm and occasional jocularity in the face of danger was a characteristic miners associated with being masculine and able to do a "man's job." Women had the same outlook. Sally told me:

> I worked with this one preacher. We were working in a pillared section and we would have to set timbers. I'd kid him all the time. I would get real close to where he was setting a timber. We was sitting and waiting for that to fall in. I called him "Preacher" and I'd tell him, "I'm going to get close to you because you're prayed up. That top ain't going to fall on me." He would always tell me, "Now honey, you ain't never prayed up." I'd always joke around with him.

Like the men, the women suffered their share of injuries. Some, like Carolyn, had "a lot of smashed fingers, broken fingers, broken thumbs, and all that," while others' injuries were more serious. Both Renee and Lynn had their teeth knocked out when working around mining machinery. "I got hit," Renee said. "It hit my lip and knocked my teeth out. I looked like a monster. My face was swollen up. My lip was swollen." Some months later they were both back at work. Even so, few of the women ever mentioned being afraid of the mine and its dangers. "I'm not scared. I have a fear of it because you know you have to," said Sadie, "but it don't bug you all the time. You have a fear, you're conscious enough to know something can happen, [but] if you let it bother you or worry you, you wouldn't go back."

Once hired, all new miners were assigned to the entry-level Grade 1 position of general inside laborer (GI). Their tasks, what miners call "brute work," included many of the physically demanding manual jobs done underground. As coal is produced, GIs perform the support and maintenance duties necessary to keep the whole operation moving. Routinely, their work consists of rock dusting mine walls to keep the coal dust down, hanging ventilation curtain, setting timbers for roof support, shoveling coal that has gobbed off the belt line, moving the belt line structures and power cables, laying track, and keeping the mine free of debris. Moreover, once or twice a year the long-wall system, or "plow," the massive machinery used to cut coal from the face, is moved. GIs manually disassemble, move, and reassemble the heavy and cumbersome pieces of equipment, such as top rollers on a belt line weighing 150 pounds and 750 feet of power cable relaying approximately 7,000 volts of electricity to heavy coal-cutting and loading machinery. Unlike miners who are classified as having operative jobs (Grades 2 through 5), GIs are given their assignments daily by their section boss. Women who are GIs and beltmen said they wouldn't know what they would be doing or where they would be sent until "they line us up outside" just before beginning their shift. Work assignments and their work location were made solely at the boss's discretion. "It's very matter of fact," said Alice, a short, sturdy single mother in her midthirties. "The boss tells you what to do. He writes it down on a piece of paper and gives it to you and you go and do it." Ellen, who was usually working on the belt line, said, "When you're general inside, they can make you do anything, like shovel a mud hole or hang rag. That's hard work." Kelly and Alice, who had worked together on the same crew for four years, commented that their first few weeks underground were hard for them. "We was hired the same day. There was about five of us. Remember?" Alice said, looking over at Kelly, who nodded. "The boss told us to get rock supports and timbers to use. Rough. It was rough for me, and [after the first few days] your body physically could not move, but you had to do it anyway. These jobs are something different and women aren't structurally built like men." But, Alice concluded, "They hired you here to work, and they expected you to do what they'd tell you to do." Ellen also talked about her difficulties putting up ventilation curtain:

> When you're hanging rag, that's the toughest job in the mines. You had to lift [and] drag three boards and two timbers. That old cloth stuff, the rag, they call it. You get real dirty, and you have to tie this and that up. I'm short anyway. It's just different stuff. All you got to depend on is the little light on your head. It seemed like nobody felt sorry for me, but I wasn't no man.

While the women readily acknowledged that their strength and stature weren't the same as the men's, as with most women in nontraditional blue-collar jobs (Deaux 1983), they insisted that they had adjusted to the work within weeks of being hired. "It was just hard work, [but] we can do it now," Alice remarked. "It's still hard work," said Kelly, "we've just adapted to the conditions. We've just gotten stronger and learned the ropes, basically. [But back then] it was a whole new world." At the time of our interviews, the women expressed satisfaction with their jobs, even though the sources of it varied. Most of them mentioned the high wages and financial security, but some women tended to be less enthusiastic about coal mining than others. Carolyn, who was a mason most of her twelve years underground, said "It's hard work, a lot of lifting all the time. I don't like it, but it's got good things, it's got bad things. It's a job. I make good money, and that's it." But for others, although the high wage was important, doing their jobs had certain intrinsic rewards, too. Of her eleven years as a GI, Alice said with pride, "I had to shovel gravel up [at the face] off onto the plow under the track. But I like my job. It's dirty, it's hard, it's cold and wet, but I like my job."

SEX-TYPING OF JOBS UNDERGROUND

Since being hired, most women at the mine had been working solo among all-male crews. While it is both legally mandated and customary for more experienced miners to watch out for inexperienced miners on the job, the women believed that, initially at least, they often captured a larger share of the men's attention. Typically, token women experience heightened visibility (Kanter 1977). The "[foreman] was having us do stuff [and] they'd [men] would stand back and watch," Alice said, shaking her head. "And laugh at you," Kelly added. "Yeah, and laugh at you," said Alice. "They put us through hell when we first get here." But, as Kelly concluded, "They work everybody pretty hard, the men, too. I'm sure they were watching to see which one of us would or would not make it, because not everyone makes it."

But before some women ever entered the mine, they were led to believe that they couldn't do the work. Kelly's grandfather, a former miner, had her shovel gravel to get a feel for what lay ahead of her as a miner. "He had a truckload of gravel," she said, "and he told me, 'You shovel this gravel out of the truck and onto the roadway and then you tell me if you want to work in the mine.'" At the time, "it was a tough job, and I thought about it. I didn't know if I wanted to or not because he said, 'This is easy compared to what you're getting into.'" Although shoveling gravel gave her doubts about ever working underground, looking back, she said, "I didn't expect the equipment and the majority of the brute work, [but] I expected [it to be] tougher than it was." Other women said that when they first started, some of the men told them that mining jobs were just too hard for any woman. Indeed, both the women and men agreed that most men had felt that way. One male miner said the bosses "didn't think they could do it," believing that since "we've got these women, we've got to get as much out of them as we can. But we've got to take care of them. Don't let them get hurt." Another miner said that "like 90 percent of the men," he believed that "they really wasn't capable of doing all the work that was required of them." Several other men insisted that men were simply stronger than women. "Well, for one thing," one said, "women just aren't built like men, and they can't lift heavy things." While he did say that there were strength differences between men, he insisted that "most women just can't do the work that a man can, like masonry where you've got to lift heavy blocks all day. A woman just can't do that." I pointed out that there were several women who had been masons during most of their tenure at the mine. After a brief but tense silence, he stated that although Kelly and Carolyn were excellent masons, they were exceptions. Generally speaking, he asserted, women just weren't strong enough. Even when faced with a glaring contradiction, some men held on to the stereotypes about women's inferior capability.

In their resistance, most women believed that the men's claims were exaggerated because if women were not able to do "brute work," then why were so many of them still working underground? They likened it to the superstition that women were bad luck in a coal mine. To them, these were men's excuses for erroneously believing that women shouldn't be coal miners. When one of Ellen's male co-workers asked, "If you can't do the job, what'd they put you up here for?" she explained that "they didn't want you to [work]; they don't want you to even try because you're crowding in on their turf." She believed that many of them still felt that way, too. Indeed, many of the women felt constant pressure to prove themselves fit for mine work. Althouse (1974) found that new miners' job-related tensions stem from worries about their own technical competence and the extent to which they can rely on others. Similarly, the women at the mine reported that all new miners felt the need to perform well by working hard, but that they had felt more pressure to do so because they were women. According to Kanter (1977), tokens' sense of proving themselves is a result of their heightened visibility. The men's extra attention only increased the pressure they felt to perform as well as if not better than the men. Ellen explained to me, "The women I have worked around, they're just as good a worker as the men or better workers because they want to show people they can do it.

Most of the men I meet, they do their job just about like I do. I mean, there's no difference." Kelly told me emphatically:

> I think I worked hard and I did the jobs I was told [to do]. So they respected me there. They didn't have to worry about a woman hanging rag today, or we have a woman shoveling belt today so help out if you can, or we're really slow today because there's a woman hanging rag or running a roof bolter or whatever. I don't think that's ever been said about me. So I think that each of us has had to prove to ourselves also that we can do the job we are in there to do.

Most regarded their jobs as a personal challenge that motivated them to work hard. This was particularly true for women who were short or slight of build. For them, proving themselves was doubly difficult because they were women and they did not appear to be "strong." Carolyn, a mason, remarked, "They [males] looked at me and said, 'You're awful little to be in coal mining.' Everybody was watching. Ah ha, that woman can't do nothing, because I was real skinny. I was really little. And you had to prove yourself, you had to prove that you're a hard worker. If there were some things you couldn't lift, you had to prove yourself another way. That's just basically the way it was." She added that many of the men "felt all we could do was shovel," even though that was "one of the hardest and most strenuous" jobs underground, but she resolved the contradiction by insisting that "you can't say that it's more physically strenuous than roof bolting or running a buggy because you have no experience. Then, if you say to the foreman, 'Why don't you have that man doing that?' well, why should you say that if you're trying to prove that you can do just as good a job in there?"

As long as some men were convinced that the women couldn't do their jobs, they believed that they would have to do those jobs for them. Some men also believed that women shouldn't do certain jobs because they were too dangerous. Either way, these beliefs only added to the men's resentment of women miners. An older male miner who had once worked with Jolene said:

> Through my raising in the Appalachian mountains, I guess, I wanted not to put her in a dangerous situation. I'd rather go do it, and I did it lots, and not for any other reason other than I

> didn't want to send her in a water hole. To be honest with you, I don't want to change. That's just the way I did things. I'd tell her, "You stay here and I'm going to do this." There was times when she would want to help me and I was afraid she was going to get hurt. I definitely did not want her to get hurt, so I chose not to allow her to do these things. I would go do it rather than be disrespectful. That's how I was raised. [But] I think I didn't help her any by doing that.

It is easy to see how this miner's traditional beliefs about gender roles led to the gender-typing of work underground, discussed in greater depth below. At the same time, many of the younger men simply expected women to "pull their own weight," and at times refused to help them as they would a man. Most women responded by doing the work themselves, without asking for help. For example, all miners were supposed to ask for help if they had to lift more than forty pounds. "They tell the men in the safety class, if you have a hard time picking up something, you have somebody help you. And it's the same thing with the women," said Carolyn. "You've got some men who will not, will almost refuse to help a woman, even though they'd help the men. The men will help each other sometimes unless you ask for help." Carolyn paused, chuckled, and added that even if she had trouble lifting something, "Naw, I wouldn't ask for help."

Other men, however, insisted on helping the women with even the easiest of tasks, which actually made it more difficult for the women to prove themselves because it put them in a double bind. On the one hand, those women who refused help, regardless of how tough bosses or men miners made their work, were viewed as acting independently, which was considered inappropriate behavior for women. One of the older male miners said with a sneer that the women "wouldn't let nobody help them do nothing. They'd chew you right out. And they've stayed here, become all independent and everything and, well, here they are." By offending their male co-workers, these independent women risked losing their cooperation completely. On the other hand, a woman who either asked for help or simply allowed a man to help her would be viewed by others, both women and men, as being either unwilling or incapable of doing the work and therefore not deserving of her job. Any woman who consistently accepted unnecessary assistance from a man ran the risk of becoming dependent on him and being resented by other miners for her apparent unwillingness to work as hard as they felt they did. "It pisses you off," said Alice, "[but] it's as much his fault as it is hers." Being viewed categorically and not as individuals, women who accepted men's unnecessary help also inadvertently reinforced men's views about all women's inability to do the work.

All of the women I interviewed noted the importance of establishing a good work reputation within the first few years of working underground. Model coal miners were typically recognized by co-workers and bosses as being able and consistently willing to work hard. "My advice to anyone going into mining," Kelly said, "is to get the toughest job underground and go at it." Miners also needed to have a good attendance record, with few if any absences. Not only did the company disapprove of absenteeism, but it created hardship for the crew. "You can be slow," said Kelly, "but you have to be there." In addition, having a bad work record or making mistakes on the job was often grounds for dismissal, and a miner who had a bad work reputation risked losing the union's support. Miners felt a great deal of responsibility to each other to work efficiently so that others wouldn't have to take up the slack. Likewise, a boss's reputation with the company depended on his crew's willingness to pull together and work cooperatively, because when the work did not get done, this reflected poorly on the boss. Crew members can influence what others, including the foreman, think of each other based on a miner's work reputation. Alice and Kelly told me, "What we have to say about each other means a lot," regardless of gender. Just as a foreman can refuse to take a miner on his crew, miners can affect his decision to do so. Alice gave me the following example:

> Like, if we're going to have a belt move, sometimes they will send us an extra person or two from someplace else. When they tell us who it is, [if] we know that person is going to go up there and lay down, all we have to do is say no and they don't ask why or nothing. Just no, we don't want that person. Get somebody else. Why send somebody up there who's going to sit there on the rib and watch you? Send somebody who'll help you, and that's what we want, somebody to help us. If you're a lazy good-for-nothing, they stick you somewhere where they can't depend on you. So the harder you work, the more they depend on you. So reputation is everything, and once you get a lazy reputation,

> no matter how hard you work from that point on, you still have that reputation.

Most of the women and some men agreed that establishing a good work reputation was harder for women than it was for men. One of the men said, "We've got some men that don't put out half of what the women do, but the union and the company can't touch them. The contract says you got to make a reasonable effort. Some guys think a reasonable effort is getting up from one seat and going on to another. It's gotten a lot of folks off the hook." But, according to the same miner, the contract didn't get women off the hook. Rather, as Carolyn said, "There was a lot of women who didn't care and didn't do anything, but then there were a lot of men who was lazy. You couldn't get them hardly to move. They couldn't say much about the women, but they did. It's awful, but it's true. A boss would make it harder on that woman and they would have taken her to the office [for reprimand]. Even if she did do a lot of work, they'd still take her in the office. It doesn't make any difference."

Both women and men had stories about recalcitrant co-workers of the opposite sex. "The men complain that the women's not carrying their weight," said a male miner. But, he added, "We've had women go to management and complain that the men wasn't carrying their weight. It works both ways." As Kelly observed, "I have seen men, big, strong, healthy men, stand back and watch *us* do all the heavy brute work. Now, that happens a lot of the time." So the women found that they had to be as assertive with other men as the men were with each other. When Lynn was roof bolting, she said:

> You just have to let them know. There's some men like this one guy I used to bolt with. The boss told us one night to go get our pin supplies. Well, he was going to sit on the back of the bolter and sleep. I kept carrying him and carrying him, and he never did come and help me. So I just made all the pins up and put them on my side. When we got ready to pin a place, he come over, and I said, "If you take one of them pins, I'll wrap it around your neck." And I cussed a little bit, and the boss got scared. He went to the miner and said, "I believe her." I don't want to see it if she's going to hit him. But you just have to put

> them in their place or they'll make it as rough on you as they can.

From the beginning, the women knew they had their work cut out for them because of their visibility as tokens and the double standard to which they were held. They knew that if the men expected less of them, they would have to give more. One of the men explained to me that if a man believed that working around women meant he was "going to do all the carrying of the blocks," but if in fact "the woman will go up there and do all the lifting, then the guy would have a different perspective. He would look differently at the women as workers." One of the bosses told Jolene, "The first day I was here, they gave me you, Kelly, Alice, and [another woman] to move power. We were riding in, and I thought to myself, 'My God, what did I do to deserve this?' Kelly [told me] to get out of her face and go to the dinner hole, sit down, and you all would move power. I felt so bad because I thought they were screwing me giving me four women. You all moved it faster than anybody I have ever seen." In the same vein, Kelly said, "[When] a couple of guys come on our section that have not worked around women, they are astonished at the things that we can do, the weight that we can lift. And if they go to get it for us, [we say,] 'No, we can get it.' Once they work with us they realize [that a woman] can pull her weight." "As the years went by it started changing," said an older male miner. "I'd say within five years, the men got pretty well adjusted to them and seen that a lot of them could pull their load." Nonetheless, doing their jobs better than some men, Jolene told me, had the following result:

> Even now, the men I work around are surprised. I will unload a flat of rock dust or something and I can outwork a lot of them. They've even told me that I've embarrassed them a couple of times. When the air changes, you build the stoppings and then you have to take a sledgehammer and knock them out. I went with one of them one night to knock these blocks out, and I was beating them out and he said, "You're embarrassing me." I knocked more blocks out than he did, and he got up there and said, "Here, let me have that [sledgehammer]." I can work with any man here, and they know it. There's enough [women] that has worked with men right now.

Kelly explained that "sometimes the men are intimidated by the women, I think, especially if she can perform her job as well as [they can] or maybe even better. I've never tried to intimidate any of the men. But you'll find a few [who are]." This was incomprehensible to her, because surely the men knew that the women were not going take their jobs. Likewise, an older male miner remarked, "If they took out all the men and required the women to do the work, I think [the mine] would fold up. I don't think they are taking the whole show. If you actually put them in some of the jobs the men have here today," he said, "there probably wouldn't be but ten of them here by next year." Because women's presence represented a challenge to the masculine identification of mining, men found it necessary to "put women in their place."

Some men believed that women could do "brute work" but thought they were unsuited to running machinery. During my interview with the superintendent, I pointed out that women at the mine were disproportionately concentrated in entry-level jobs such as GI and beltman. He maintained that the men had more experience with machinery, had a better understanding of how machines worked, and took "a more mechanical approach" than the women. He explained that the women's traditionally gendered socialization had handicapped them for work in the coal industry. They remained in the less-skilled jobs because of "the natural settling of their skills and their application." When gender bias is condoned and legitimized by top-level authorities, it becomes pervasive and entrenched (Reskin and Padvic 1988). The women told me that the men, usually the foremen, lacked confidence in the women's abilities to perform even the simplest of mechanical tasks. According to Darlene, foremen thought that "women are harder to train. You know, like we're dumb or something." Sadie said with a note of bitterness that the men she had worked around "don't think women are smart enough to put something together, which I can do. I've done a whole lot. And the boss goes right along with it. It's hard [when] he's got the men putting [machinery] together. I don't think it's right." Usually, foremen simply ignored women when assigning operative tasks to miners. Darlene, who had advanced to wireman (Grade 2), recalled that during her crew's dinner break a boss called on the phone and asked one of the men to check on some wiring. Realizing that the task was really her responsibility, she took the phone and said to the boss, "Yoo-hoo, I'm sitting right up here. I can do that for you." The boss paused and then, according to her, hastily explained that he had called on this man because he had his tools with him. The truth of the matter was that neither one of them had their tools with them. From that day on, she vowed never to be without her tools. Carolyn told me a similar story:

> I had this one boss, and he just bypassed me on a job he knew I could do for another guy who never even run a motor. That kind of made me feel bad. He just looked at me and went on. I've been on a motor moving it around all the time. Taking it in and out wasn't a problem. The boy that I work with all the time just looked at me after we got around to the other side and started laughing. He understood what was going on. Most of the men did. We just sort of looked at each other and said, "Can you believe he did that? Boy! Is he a chauvinist!"

Other women who had similar experiences confronted the bosses about denying them the opportunity to run machinery. For example, Kelly, who often worked overtime, had asked her boss if she could come in early one day and was told he had no work for her to do. Later, she found out that three men had come in early that day to do work using a jackhammer. "I said, '[Foreman], I thought you didn't have any work for me to do.' He said, 'Well, I don't.' So I told him I can run that jackhammer as good as that man can. Then we really got into a shout down, scream out. I was so mad I walked off. I cried I was so mad. [But] from that day on [the foreman] treated me as an equal." Lynn was also effective in thwarting the efforts of a foreman who routinely denied women work assignments running machinery. She had been a roof bolter but was reassigned to GI when she had the following experience.

> We've had a couple bosses up there that thought women could do nothing but shovel. I had one foreman [who] had me on a section as an extra person to hang rag. I roof bolted before that, but I got called back to GI and roof bolters would be off. He would send the other [male] GIs that was younger than me to roof bolt. Well, I went to the union to file a grievance on it, and the union talked to him. After that night I roof bolted until they sent me to [another shift]. If it hadn't been for me going to the union, he would have kept doing me that way. It's just the type of person he was. Everybody knew that he didn't think a woman can do nothing but shovel. He was one of those bosses that was

> like that. You have to stand up for your own rights. If the women don't stand up, they will do you that way.

In their overview of gender inequality at work, Reskin and Padavic (1988) found that employers generally give men rather than women workers the kinds of tools that enhance men's productivity. Moreover, Cynthia Cockburn (1985) has noted the gendered nature of tools and machinery in the workplace. Machines and men are naturally correlated and are off limits to women. This assumption is fundamental to those hegemonic processes operating in virtually every male-identified work setting in which men have a stake in preserving their dominance. I learned of this dynamic in the context of the existing power structure between a boss and his crew. "They don't really want us here in the first place," said Michelle. "So I think that most of the bosses depend primarily on the men. The men are bigger and stronger than we are, so naturally they would want the stronger workers on it." However, most women agreed that having less physical strength and endurance than men would affect job retention rather than advancement to a higher-grade job. They pointed out that some women who started at the mine with them quit within weeks of being hired because they lacked the stamina to do the work. Acceptance and recognition by male co-workers and supervisors was more important in affecting women's advancement in occupations, such as coal mining, that have had strong male-identified traditions for work and social relations (Deaux 1983). As noted above, a miner's work reputation was important not only for being respected and getting along with other miners but also for gaining the kinds of opportunities necessary for being promoted to higher-ranking jobs. Conversely, if a miner had a poor work reputation, there were ways to deny her or him a promotion. To the extent that women must work harder and have better work records in order to get these opportunities, they were at a disadvantage. "Men have it in their minds that we are the weaker sex. I think the bosses think the men can move the buggy faster, they can do the work faster and better than us," said Michelle. "But that's not true." Ellen agreed, adding that with "a piece of machinery, the women are faster than the men because they want to do it good. Like Lynn can roof bolt real good."

Despite the apparent discrepancy between some men's stereotypical perceptions and the realities of women's abilities, bosses, with the support of some male miners, regularly assigned women to work that reflected women's traditional gender roles. This sex-typing of jobs begins with sex stereotyping, apparent in the controlling image of the incompetent sexualized woman miner. Foremen rationalized the clear sex bias in job assignments in terms of the women's suitability to jobs requiring the traits women are presumed to possess on the basis of their "essential nature." Male miners and managers told me repeatedly that, despite women's limitations, "there are *some* jobs women can do in the mines!" One man claimed that "women are most suited for gang work. They've got men working with them. A lot of women work the belts. A lot of women are masons. Something that involves gang work where they've got a man there to do the heavy lifting and the women do what they can." When I asked Ellen if there were "women's jobs" underground, she responded, "Oh yeah! You got yourself on the belt, that's a woman's job. You go shovel the belt, you help the mason build the stoppings. Most of the time, if I'm not by myself, I am helping one person." Sally agreed, adding:

> We would go on a section and you'd have a smart aleck call you "girls." I have been called a "hen." [And] this one guy would say, "Send the hen down there to do that, or send the girl to do that." To me the equipment was safer than where they sent the girls, like set timbers or shovel the belts. It's more likely you could get tangled up in that belt than you would pinning top or scooping. What the women would go do would be the hardest work. The men didn't want to have to do it. It's like they're doing the women a favor, they don't need to be on equipment. They send the women to do the dirty jobs.

Not only were women presumed to be incapable of doing newly redefined masculine types of work, they were often expected to perform tasks that mirrored the work they traditionally performed in their homes in support of and in service to men. As with the "housewifization" of women under patriarchal capitalism, women miners were subject to the same ideological processes at work, with the same results. For example, Sadie likened her self to a "mining housewife," saying:

> Sunday I carried cinder block behind them, I carried rockdust, I put belt together, I cleaned up the garbage, I carried their junk to them if they wanted to work with it. It's just like you're a

> gofer or something. Go get me this, go get me that. You just do all the tasks that they don't want to do. Like the men, when they set up a belt head, they throw down everything. They never put nothing away. Well, it's up to us to go clean up their mess.

And as Carolyn said irritably:

> I'll tell you one thing that really bugs me. When I was general inside hanging rag, it didn't matter what section I went to, they'd expect me to clean the dinner hole and then go out there and hang that rag. That really upset me. And I would tell the boss, "I have to do this job because you tell me to, but I want to tell you right now, I resent this because my trash is in my bucket and that's where their trash ought to be." It's just like—there's a woman, let her clean up the dinner hole. Now, that really bothered me, bugged me to no end.

During our 1996 interview, Carolyn again readily acknowledged that the men continued to "give the women certain jobs and the men the ones with the responsibility. Because they are used to their wives working around the house, they think that the women can do that kind of work, too." Some women also said that that certain jobs, such as general inside labor and beltman, were stigmatized among underground workers. More highly classified men, said Carolyn, "thought if you were general inside you were low." According to Ellen, "[As a GI] you're the flunky. I mean, you're the gofer. It's real hard. Like yesterday, me and two GIs, the two men that work with me, had to shovel out a big old hole in the ground where the boom's at all day." Sadie added, "They think you're stupid. Most of them GIs, you go do this and you go do that. I know all the women experience work discrimination because most of us are gofers, hard manual labor." The work that required the greatest physical endurance, the least skill, and was often the most dangerous was considered "women's work." Darlene said she suspected that the company wanted women "to be GIs and not move up at all," so that ultimately they would get discouraged and physically worn out and would quit and go home. Ellen concurred, saying, "It would be hard to boss a man up there because they wouldn't listen to you very much and they probably don't really like us up there anyway." Thus, over time, the men's persistent stereotypes became the justification for restricting the women as competitors for better-paying, more prestigious jobs by redefining the sexes' respective places in the work hierarchy. As the following discussion shows, the formal mechanisms for this process further legitimized the gendered division of labor underground.

ADVANCEMENT AND THE GENDERED DIVISION OF LABOR

Several years after women miners began working in greater numbers in the nation's coalfields, they voiced their concern that they were not advancing to the better-paying and more skilled jobs underground. At their first national conference in 1979, women miners complained that they were often relegated to the more menial and lower-paying jobs because they lacked the opportunity to get the on-the-job training necessary to advance (*Business Week* 1979; *Mountain Life and Work* 1979). During the mid-1980s advocates of women miners nationwide told the press that women were given the heaviest, dirtiest entry-level jobs in the mines and that they stayed there long after men would have been promoted to more prestigious and better-paying positions operating machinery (*Coal Age* 1986; Keerdoja, Foltz, and Barnathan 1982). Only a few studies of women coal miners have examined women's advancement and the resulting job-level sex segregation. According to a U.S. Equal Employment Opportunity Commission study (1979), 57 percent of all female underground workers in 1978 were general inside laborers, compared to 20 percent of all men. The same year, Mahoney (1978) described the entry-level experiences of fifteen women coal miners in two Appalachian states. The women reported having difficulty getting the opportunity to learn how to operate mining machinery so they could be eligible for promotion. In a Coal Employment Project survey done in 1981, 36 percent of the female respondents said they believed they had been given "harder, more dangerous, or less rewarding jobs than their male co-workers," while 47 percent said "they would have gotten better jobs inside the mines by now, had they been men" (White, Angle, and Moore 1981, 8). Based on interviews with both women and men miners in three western states, Yount (1986, 676) concluded that "women miners are not advancing at a rate commensurate with male co-workers. That is, a division of labor by gender has risen within coal mines in which the majority of women are segregated into low-prestige laborer positions." Based on my analysis of a 1986 Bureau of Mines data set, women were disproportionately segregated in the lowest-ranking jobs in the nation's coal mines relative to men (Tallichet 1991). Compared with miners nationwide, men at the mine I studied were similarly distributed across job grades, while women at the mine were even more disproportionately represented in Grade 1 jobs, such as GI, beltman, and mason. (For a complete listing of jobs underground, see the Appendix.)

For the women at the mine who wanted to advance, the formal contractual procedures governing this process often served as formidable barriers, although not impenetrable ones. These formal barriers had not been operating in isolation of the informal processes by which men devalued women at the mine. Women's physical endurance and technical capabilities were always in doubt. Machinery, as masculine-identified objects defining "men's work," constituted the basis for task differentiation between the sexes underground. Task differentiation between women and men miners reflected the social differentiation of the localized subculture and its corresponding ideology, as previously discussed. Moreover, organizational practices meant to ensure fair and equal treatment of workers on the basis of merit were used and misused to provide men with promotional advantages over women. As a result, men had advanced based on their real as well as their perceived qualifications, while most women had remained in entry-level positions. By understanding the women's responses to their work and workplace, it is possible to document how a pattern of gender segregation had begun to emerge at the mine.

Formal Barriers to Advancement

Two major problem areas for women's advancement were on-the-job training and seniority. While job certification as a means of job entitlement was a matter of successfully completing a classroom course, on-the-job training proved to be more problematic for the women at the mine. According to Sadie, she and the other pioneers were simply denied the opportunity to get task training. A company policy that created opportunities for women to get trained on machinery went into effect in the late 1980s, but most of the women said they were rarely given the opportunity to use their skills. Moreover, when women were temporarily assigned to positions operating machinery, their experiences did not lead to a promotion because they were not retained in these positions long enough to petition to keep them.

For UMWA coal miners, seniority is defined on the basis of years spent working at a particular mine and on having the ability to step in and perform the job. Their inability to learn or use operative skills meant that some women were simply not qualified either to fill these jobs temporarily or to bid on them even if they had the necessary seniority in years. Moreover, when women did bid on more highly classified jobs for which they were qualified, they often had to prove their ability to do the work, something not required of men. Most often, the women felt they did not have enough years at the mine to bid on jobs. Although this barrier was more perceptual than real, it contributed to the women's unwillingness to bid on vacancies as they were posted. Eventually, owing to the coal industry's slump in the early to mid-1980s, job vacancies declined sharply. Moreover, when jobs were posted, women who were single parents were unable to bid on them because of their childcare responsibilities. In this way at least some women were more disadvantaged than men when competing for highly classified jobs underground. Finally, periodic realignments of jobs resulting in demotions for miners with less seniority have also more disproportionately disadvantaged women. As realignments occurred, women who had advanced were demoted from their more highly classified jobs. Let us now look more closely at women's experiences with job certification, on-the-job training, temporary assignments, seniority and the posting and bidding system, and realignments.

Job Certification

One way to qualify for a classified position underground was to complete a course that certified a worker for a particular job. In 1983, for example, Kelly had successfully completed a course to get a shot fireman's certificate qualifying her to do this Grade 3 job. "Training's easy to get," she said. "On a Friday, they said, 'Come for a Saturday class.' They gave us test materials to study. You take your shot fireman's test, and you are a shot fireman. Every single person in that coal mine has had that opportunity. I was the only woman in the class. It was great. I loved it. I made the highest score." Among the women, the most popular certification was for the Grade 5 job of fireboss. This training consisted of an intensive three-month course during which miners were required by state regulation to pass each section of the mine foreman's exam with a score of eighty or higher. Miners who had firebossing certification and the requisite seniority in years could be called on to be fireboss during their shift as needed. "The fireboss goes throughout areas of the mine," Kelly explained to me. "They check for bad top, gas, any violation. You fill out forms and it's got to be right. A violation could be a roof bolt that's messed up. You got to set a timber under it. You got to shovel the belt, rock dust the belt, pump water, and things like that. You write those up and someone has to correct the situation." In essence, when a fireboss finds a violation, miners have more work to do. "I have a pretty important job being a fireboss," said Jolene, "[because] the fireboss checks the line. I catch flack sometimes. You find a damaged bolt and they have to come back and put another one in, or there's something on the track that needs to be moved. It needs to be done." But more than that, said a veteran male miner, "As a fireboss you have a lot of power. You can bring them [the company] to their knees, halt operations."

In 1990, Jolene, Carolyn, and Jessie became certified as firebosses. Kelly and Ellen had considered getting this certification but were apprehensive about passing the test and handling the responsibilities. Some women had refused to get this certification because of the added responsibility. "They can make you fireboss, walk forty breaks of belt," said Carla. "They give you a direct order, you got to do it. Then if something would happen and you would overlook stuff, which is normal for anybody that could do it, somebody comes behind you. They take you in [to the company office] and jump on you for it." But by 1995, Kelly, Ellen, and Alice had taken the course, passed the tests, and had their "bossing papers." The superintendent told me that the class had been open to all miners and that "a larger percentage of women than men had taken advantage of it." These six women constituted roughly one-third of all miners at the mine who were certified to be fireboss. The women's disproportionate representation led a male miner to remark, "It's easier [for a woman to advance]. I think we have twenty-five women here. Five of them right now are doing fireboss work." Despite his perception, Kelly told me, this form of advancement had not come easily to any of them. "The reason I didn't take it [fire-bossing class] earlier was the fear of failure," she said. "It was my reason for years, [because] we're all afraid of failing in our brothers' eyes, and when you do, they let you know you messed up. Some men won't take the bossing class either, but I think that's a lot of our reasoning. So we tend to shy away from things we're afraid we'll fail at. I was scared to death, [but] I had to quit being afraid of failure." As we have seen, most miners worry about their technical competence and how it will affect the well-being of other miners. For women miners, the pressure was particularly acute as they struggled to get past the men's stereotypes. For years, Kelly and Ellen had wrestled with self-doubt and were haunted by the prospect of failing to prove themselves. But conquering this self-doubt was essential if a miner was to persevere. Ellen gained a sense of her own empowerment upon realizing the false nature of patriarchal ideology. She felt that the men with whom she worked simply were not superior to her in ways she had once presumed them to be. Thus, she overcame her apprehension about getting her bossing papers because, as she said with a guffaw, "You see these men that's got less time than you, and you think, why, if they can pass them papers, anybody could."

But if the process of getting the certification provoked anxiety, the added pressures of the job were even more stressful for those women willing to take such a chance. Doing the job itself, said Ellen, "worries me because I am a woman, and if there's something I overlook, an inspector comes over and would tear my butt up or somebody would get hurt or killed [because of] something I had overlooked." Even so, the higher pay and the job security made this job worth the added pressure and risk. "You better get all you can get, because that mine ain't going to last," she explained. A miner with more skills had a better chance of being called back after a layoff. Generally speaking, miners with more years of seniority were called back first, but their qualifications more often determined how quickly they returned to work. So it was to the advantage of less senior miners—namely, women—to become as qualified as possible in order to avoid being laid off indefinitely. "What every woman should have done [was] learned from the beginning how to operate every piece of machinery in that mine," said Lynn. "If you know it all, then you've got a better chance than any of them of staying. So I just made up my mind that I'd work and do the best I could, and the more I could learn, the better off I'd be." "We know the mines is going to finish up in ten years or less," said Kelly. Then the only jobs available would be in the small "punch mines." "Men are lucky in a sense, [because] I don't think small mines is going to hire a woman," she said. "I don't for lots of reasons, and there don't have to be one."

On-the-Job Training

According to federal law, inexperienced miners are considered trainees during their first few months underground.[1] Their trainee status is des-

1. Originally, miners were trained on an apprenticeship system. After the mines began to be mechanized beginning in the 1920s, and for the next four decades or so, miners'

ignated by the bright (usually red or orange) color of their hard hats, and they are referred to as "red caps." In keeping with the 1988 and 1993 BCOA/UMWA contracts, red caps are forbidden to operate machines at the face and transportation equipment, to work out of sight and sound of another experienced miner, or to bid on another job until they have worked a total of forty-five days underground. Upon successful completion of their training, these new miners receive their federal Mine Safety and Health Administration (MSHA) and state certification or mining papers, which qualify them to begin working underground as "black caps." Typically, some jobs at any mine above Grade 1 are regarded as the "property" of more experienced senior miners (Althouse 1974). So, while production workers' assignments are more or less permanent, general inside laborers are temporarily assigned either to a crew or to work in the back areas of the section being mined. They have no designated task specialties or work areas, nor do they stay in one part of the mine long enough to become familiar with it or with those who work there. Thus, they have fewer opportunities for engaging in the informal social relations necessary to learn new job skills (Wardwell, Vaught, and Smith 1985). Although foremen are supposed to provide opportunities for workers to learn how to operate machinery, co-workers often assume this responsibility. Unofficially, "job switching" is common (Althouse 1974; Ross 1974). Miners are taught how to run a piece of machinery by a more experienced miner, either during periods of rapid production or during slack periods underground. Although job switching occurs on a random basis, teaching and learning different jobs and customary "tricks of the trade" occupy a considerable amount of miners' time on the job (Althouse 1974, 14).

In general, the skill level among the women in my sample varied according to how long they'd been working, partially reflecting a change in company policy during the postpioneer period. While women who began working at the mine after the gender barrier had been breached said they had no trouble getting on-the-job training, women who entered

---

training was based on a "buddy system." However, since the Federal Coal Mine Health and Safety Act of 1969 and its 1977 amendments, coal companies have been required to provide all entrants into underground mining with forty hours of safety and health training. This includes eight hours of training and orientation at the actual worksite. Some states, among them West Virginia, have enacted laws requiring additional new miner training of up to forty additional hours. Thus, in order to accommodate prospective entrants, an increasing number of colleges and vocational schools have been offering basic mining courses for new miners and managerial courses for miners who want to "go company."

during the pioneer period complained that they got no training at all. In fact, Sadie, one of first women hired to work underground, actually spent her first few weeks aboveground cleaning the company's offices and equipment warehouse. "They kept us outside with a lot to do," she said. "Housework. They don't do it now. They can't do now. They got jobs posted and men's got them. It used to be every time the warehouse needed to be cleaned, they'd want me to do it. They said I did a better job than anybody that done it. But it stopped. I didn't like staying out. That's not what we were there for." Some weeks later, the company assigned them to GI jobs underground, not because management feared legal action from the women but because some senior male miners, who preferred to work aboveground before their retirement, complained that they had not been given that option. Once underground, "when you're a general inside trainee, they show you lots of little things," several women said, but red hats "do not touch this equipment." However, this rule had been violated on occasion, sometimes for reasons of convenience. As Sadie recalled, "When I was a red cap, one guy let me run a portabus while we were bonding track. And I was moving the bus up and down the track for him. He said, 'Don't you tell nobody I let you run that bus, you're not supposed to be on it.' [But] I knew that a lot of the men that came in as red caps ended up on sections around equipment. They got their black cap, [and] they could take over. So you know they had to train them as red caps." When she became a black cap, she admitted with a note of bitterness, she was denied the opportunity to run any machinery:

> After I made my miner's certificate, I worked on the conveyor belts. But mostly they put us on belt heads back then. They don't do it now. But when we started we were put on junction points by ourselves, watching a junction point, or on belt heads, loading the coal into the cars. But the women that were there first really didn't get an opportunity to go into the face and learn to operate equipment. It never came up because we were kept out of the face. We went to the face every now and then to hang rag. They didn't offer us any chance to run any equipment.

She went on to explain that without the training, neither she nor the other women like her could ever bid successfully on a more highly classified job at the mine.

During the early 1980s, as the postpioneer women began to accrue some seniority, they began to pressure the company to get on-the-job training. According to a veteran male miner who was also a high-ranking union official, although some women did not complain because they feared reprisals from their male co-workers, others did, quite successfully. He told me that "grievances [were filed] giving the women the opportunity over men to be trained first. I think some of the foremen, through ignorance, put theirself in a real bad situation," he said, "[because] some women actually wanted to [train] at some point and they just burned them."[2] For example, Darlene, who had a good work reputation, had asked her boss to train her on some machinery. Instead, he trained three men ahead of her, all of whom had less seniority. Although she complained, nothing was ever done. In Michelle's opinion, "I think [if] a person's in a coal mine ten years, they should know how to run everything in it. They should put them on a down section and let them learn. Put them in coal [production area] and let them practice. Only one or two of us [women] have gotten a chance to do that." When she first started hanging rag on a section, Michelle, who now had a Grade 3 job, said, "You might find some guys that say, 'Come up and put up a pin.' [But] you don't find too many of them that do that, [maybe] one in two that want to teach you a little bit." Thus, during the early part of the postpioneer period, getting on-the-job training wasn't impossible, but it was harder for women than for men.

A few years before I conducted these interviews, the union, with the company's cooperation, instituted a new policy for miners seeking on-

2. According to the 1988 National Bituminous Coal Wage Agreements (UMWA/BCOA 1988, 40–41), should a work-related conflict arise between a miner and the company, the miner with the complaint should notify the foreman or "boss." The foreman has twenty-four hours to respond. If no agreement is reached by that time, the miner has up to ten days to file a standard grievance form. Within five working days, management and the mine committee must meet to settle the issue. The mine committee has the authority to act on the aggrieved miner's behalf. The mine committee consists of three to five miners elected to represent their co-workers so that each one is present on each shift. If they cannot reach a settlement, then these parties must file a grievance form that is taken up by the UMWA and a company representative in the union's district. These parties must meet within seven working days to resolve the complaint. At this time, although mine committee members may be present, those arbitrating may not be the same individuals involved in the earlier stages of the negotiations. If no resolution is reached at this point, the case is referred to and decided by a district arbitrator chosen by the union and the company within ten calendar days. Within fifteen days, the district arbitrator holds a hearing on the matter and after thirty days renders a decision. The aggrieved miner has the right to be present at the meetings during each stage of the process.

the-job training. Miners who wanted to be trained on a piece of equipment were required to sign a waiting list. While some of the women felt that this seemed fair, it gave the company greater control over on-the-job training and usurped the miners' prerogative for providing that training. According to one of the older male miners, "We probably didn't train them well when we didn't look at them as a fellow worker." Because the men were pushed into training women, he added, the company did not "require the same of [the women]. There were a lot of standards we had to abandon." Thus, even when women were trained to advance, their abilities were still in question. At the very least, the women asserted, the conditions under which women and men learned new skills were the same. Several of them reported that getting trained on equipment was easier at that point. "Basically anybody can be qualified for a higher-paying job," said Lynn. "All you got to do is go in and ask to be trained, and they have to train you." It was simply a matter of asking for it at the right time. "If things are running good, you can't get to run the machines," Jessie said. "But if things are slow, you can get to put a pin in." The women I interviewed also said that a worker's reputation, especially the willingness to work hard, often determined how swiftly his or her request for training would be granted. As noted earlier, getting a good work reputation was harder for women than for men. Once again, because of the prevailing stereotypes about them, women had to work harder to get the training they wanted. About half of the women I interviewed had taken advantage of these training opportunities and knew how to run machinery. For those who had not, it was simply not worth the extra effort. "These women don't want to push the issue that I want to be trained," Renee said. "I say, don't take no for an answer. The worst thing you can do is give up. I learned that." The women's general feeling was that if women who avoided training did not know how to run something, it was their own fault.

But the consequences for women who had been trained to run machinery were not always rewarding. Some men obeyed the letter of the policy but defied the spirit of it by making training uncomfortable or difficult for women. While he was teaching her to run a roof bolter, a male miner told Kelly, "If I teach you my job, you'll end up taking it someday." Whenever she asked for training, "there was little slurry remarks like [that]," she said. "That doesn't take place now, but it did [then]." She passed the miner's remark off as "a joke," but it does reveal that when women acquired skills, some men were threatened by the loss of privilege and power this meant for them. Darlene and Carla thought that their training experiences, which led to more highly classified jobs for both, were originally intended to punish them for aspiring to advance. Carla explained:

> I actually think I was being punished when they put me on the plow, because you could not hire people at that time to go down that jack line. They needed a jack setter, so they took me on the line and said, "All right, you set them." I said I don't know how, and they showed me in about five seconds, then left me to do it. At the same time, there was a guy there and the boss took [his] time with the guy. He stayed with him for about a half an hour.

Shortly thereafter, the machinery she was operating malfunctioned, and, she said, "The boss came down and jumped on me, [asking,] 'Why didn't you pull these jacks up, they're out of stroke?' Heck, I didn't know it was out of stroke; I didn't know anything about them." She concluded, "I got picked on, [but] once I got down there and got used to the job, I said, well shoot, I can handle this. This is good pay. I'll stay. Then I bidded for the job when they tried to take me off. Now they can't stand it because I won't quit and give it up [so they can] put it up to bid for some of the men because they got a buddy that rides with them, and they need him" on their section. Jessie, who had been demoted from a classified operative job to GI, was ordered by her foreman to train male GIs to operate the boom. Eventually she was reassigned as a GI on the same section, while one of the men she had trained was assigned her former position. In another instance, Irene, an older woman, and a young black man had been trained to be mechanics, a Grade 3 job. Although the training qualified them to do the job, it also prevented them from ever advancing to a Grade 4 job. As Irene told me:

> I would have been four if I had got certified. But they had me greasing and putting lights in, things like that. I spent seven weeks over there [in training]. Me and [the young male miner] went over there together. He's a black boy, and they had a lot of white men there. Me and [the young miner] were the only two odd colors there. He was a black and I was a woman. We bidded to get to go as trainees and they sent us to school. They had us

> tearing rock dusters apart or washing down buggies. They didn't have us training, wiring up panels and doing things like that. Now, I think [the boss] was the cause of that. I think [he] had a problem because he was black [and] he didn't want him as a mechanic. I think my part was [being] a woman.

Once women acquired a new skill, they had to "use it or lose it." But given their multiple responsibilities as GIs, they did not often get an opportunity to use the operative skills they had learned. "You don't get the experience, because if you ever hang curtain in a coal mine, you are behind when you come in," said Ellen. Although she knew how to run some of the machinery underground, she was rarely granted a job assignment that would require its use because she was usually on a "down" section, doing GI work, rather than on one "running coal." Without practice she felt she lacked the proficiency and thus the confidence to do a good job. "I can roof bolt and I can run a buggy," she said, "but if you don't do that regular, you're not going to just jump on them machines and do everything. I never went on a [coal-producing] section where they said, 'Well, Ellen, you run the scoop, and so-and-so, you go in the face and hang curtain.' It's always: 'Ellen, you go hang that curtain and so-and-so gets the scoop.'" The other person was always a man, in Ellen's opinion because "the men basically notice other men, you know. They're all pals. They don't want the flunky work. Now, if it was me and that man, I'd get the worse job." As Sadie put it, "They primarily want the men to have the buggy jobs, the roof-bolting jobs, and the miner jobs, which is still the way it is today."

Like Ellen, several women admitted they lacked the confidence to run machinery. This is fairly common among blue-collar women in male-dominated workplaces (Deaux 1983). As marginalized members of the underground workforce, most of the women preferred to stay in entry-level jobs, where they could continue to demonstrate their relative competence. Sadie, the pioneer who was denied training opportunities, declared, "I don't know how to [run machinery] today and I really don't care. I like my job." Her approach was to "stay where you're at, and you really know what you're doing." But even some women who had the training had a similar attitude. Kelly explained that "sometimes a general inside labor job, it's not easy. But there's no pressure, there's no major head-busting decisions to make, somebody else tells you what to do, somebody else takes the blame if it does not get done right. It's easy to fall into a situation where I just go about my eight hours and get it done." And, she added, "If you don't advance, you don't take a chance on being wrong or messing up, [because] when you make a mistake, they really don't let you live it down." The women in particular, she concluded, were less likely to take a chance on getting into a more skilled job underground "because we're women and we're feeling inferior anyway." Likewise, one of the men said, "A woman just can't hustle like a man can. They're not made to stand up under the stress of intimate danger of working underground like a man can. A woman has to be more cautious. I'd say they are more afraid to take a chance than we are." Women have not advanced like the men, he added, "because a lot of [them] know their capabilities and that's good, because if they try something and they're scared to death of it, they are more likely to get hurt. So if they know their limitations and if they are satisfied with eight hours of shoveling belt, that's what they should do."

Some women resisted their marginalized status, however, and established their own definition of self as worker. Sally interpreted women's cautiousness in a more positive way. "You never hear of a woman getting killed in the mines. Hurt, but never killed. Anytime there is a death, it's always a man. I believe a man's got this macho image, or they just don't think." To most men, being cautious was an "unmanly" quality that got in the way of doing the masculine-identified work of operating machinery underground. Ironically, the masculine identification of mining that forms the basis men's solidarity has been used by management to "divide and motivate miners" within the system of patriarchal capitalism (Yarrow 1985, 41). Lynn, a classified roof bolter, said, "Most jobs like that [Grade 1], you don't have a boss standing over you all the time. You just go on and do your work. Some people seem like they can't hardly handle pressure with the bosses around them all the time. I figure that's why they like it. I've done that work and sometimes you don't never see a boss except before you go in." But when she roof bolts "on a section, a boss is with you at all times, which I guess they have to be on account of the safety part, and if an inspector comes in you got to be doing everything just right. They'd just rather be away from pressure." She continued:

> There's not very many of the women up there that likes to run coal. Most of them would rather be on belts or something like that. The way I see it is mostly they're just satisfied with the

> lower jobs. Most belt work is pretty hard, I think—I mean, you're lifting and tugging on structure and belt and shoveling and that's mostly the extent of it. I like something you can learn on, like section work, where you're producing coal and you're always learning different ways and better ways and it just seems like you get more qualified at it.

Then, she said, chuckling softly, "I just like to run coal. Most of the women up there [have] settled to the kind of jobs that they want, and if they don't want to do that kind of work, they'll stay away from it. Most of the men don't care. Most of them have had a lot to do with everything anyway. It seems that the men are more wanting to learn to do everything than the women do. Some of the women say they just don't want to work in the face, they just don't want to run no equipment. They'll tell anybody that." Similarly, one of the men said, "The more abilities a person has, the more respect they are going to get from the other people." Grinning, he added, "I work on the long wall. It's the most hazardous job in the mine and I love it. I really do. A lot of guys won't go up there, and the women are scared to death of it."

Perhaps some were, but not all of them. Over the years, several women had classified jobs working on the plow at the face, where coal is sheared from the seam. One of them, Carla, said, "I think women have come a long way to prove to these men that we can do the job that they can do, especially the women coal miners. Just like me when I went to the plow. I had to prove myself a jack setter. I had to prove to the people that I worked with because it had been all men up until that point. I had to prove to the men I could do it, I had to prove to the boss I could do it." Just as the men's persistent stereotyping of women as incapable of brute work justified their early attempts to exclude women from the workplace, their equally pervasive stereotypes of women as incapable operatives affected the women's self-definition as coal miners and their aspirations for advancement. But some women had begun to redefine themselves and to resist the men's patriarchal beliefs.

In short, inexperienced miners, both female and male, begin working underground in entry-level jobs. However, by virtue of their initial physical or social isolation underground and other inequities in their early training experiences, most women did not acquire the skills that would qualify them for more highly classified jobs. As a result, men with equal seniority advanced faster than most women. Some women also became discouraged by repeated unsuccessful attempts to get training and simply gave up, while others never bothered to try. As Ellen allowed, "I think they just accept theirself in that position. They like it, they don't like it, but they're there, and they're afraid to advance theirself." Because men are assumed to be the "natural inhabitants" of work organizations, their power and the legitimacy of their definitions of others are rarely challenged (Kanter 1976). Furthermore, following Kanter's (1977) analysis of tokens, Fonow (2003, 82) has stated that "it may take [women] more time to untangle mistaken identities and to establish competence-based working relationships, particularly with the dominant category." Once these working relationships are built, women are reluctant to change jobs. In this way women can avoid conflict, minimize gender-based assumptions about their competence, and blend in with members of the dominant group. According to Collins (1990), this is typical among subordinate group members as they develop a dual consciousness when coping with oppression. At that point, the underlying motivation for "staying where you're at" was to reduce their visibility and minimize the men's everyday devaluation of them.

Temporary Assignments

Under the contract, foremen could temporarily assign a qualified miner to a more highly classified and higher-paying job for as long as he deemed necessary.[3] For example, a miner permanently classified a GI (Grade 1) who also had a fireboss (Grade 5) certification could be asked to act as fireboss for a given period of time during which the miner would receive the higher rate of pay. Also by contract, if the miner worked in this temporary capacity for thirty or more consecutive days, then she or he could ask to have the job posted for anyone to bid on. Usually the same miner bid on and got the permanent position. In the event that the job vacancy remained unfilled past the time required for its official posting, foremen could select a miner to be trained for it. The intent of this procedure was to "to help senior employees achieve

3. In both the 1988 and 1993 National Bituminous Coal Wage Agreements (UMWA/BCOA 1988, 1993), Article XIX, sections b and c, state, respectively, that within sixty days employees must be "classified in a regular, recognized occupation. Failure to so classify . . . will result in automatic classification at the rate which is the highest rate for any work performed," and that "where a senior Employee has expressed a desire to improve his ability to perform a job to which he wishes to be promoted, to the extent practicable, he shall be given a preference in filling temporary assignments in regard to that job."

promotion," and so "they shall be given preference to the extent practicable in the filling of temporary vacancies" (UMWA/BCOA 1988, 97). But the contract also states that management retained the right to deny a miner the promotion. The only recourse a miner had then was to file a grievance with the mine committee against the company. But a veteran male miner and former union official told me that, since most grievances had already been litigated, miners rarely won. Thus, foremen dictated who got temporary assignments and ultimately made the final decisions about who was promoted.

Several women said management had been taking advantage of the temporary-assignment policy for two reasons. On the one hand, management could get the work done and still cut its labor costs by temporarily assigning a miner to a more highly classified job for fewer than thirty days. Although the miner enjoyed the short-lived windfall of a higher wage, the company avoided paying the higher wage permanently. On the other hand, a foreman who wanted a particular worker on his crew could assign a miner the temporary job for the required time, have the job posted, and give the miner the job. The women I interviewed said that men were given preference over women in the assignment of temporary jobs, sometimes in order to promote them permanently. Sadie explained:

> The bosses will give the top-paying jobs to the men. It happens more often to the men because we don't usually get put on a job. We get one job and that's ours. Like my belt job. I hardly ever do anything that would need classified. Anything else I do would be GI work. That's the way it is with the women. Now, the men, if you get one of their favorites, if they want him to have the job, they're not going to jerk him off in twenty-eight days or whatever. They're going to fix it so he can stay right there. He can be classified for the job. Your women and your least-favored men will be jerked off.

In another instance, the policy was simply violated to benefit men. Ellen told me about a black male miner who was retained in an operative job that was eventually given to him permanently. However, a white woman miner had been temporarily assigned well over the thirty-day period as a fireboss without any such promise. She said that when the woman seeking permanent fireboss classification objected, her foreman told her, "'If you file a grievance I'll just take you off fire-bossing before I'll give you the job.' But yet we got this black guy [who] was a GI, and he worked on a motor so many days out of so many months. They knowed that would stir up so much stink because he was black, they give him that job. That's a real different story." Despite her foreman's threat, the woman filed a grievance and lost. In another instance of gender-based favoritism, Michelle told me, "There's a guy on my section now [who] ran a motor for thirty days. He applied to be awarded that job and they gave it to him instead of it being put on the board for bid," she said. "They awarded him that job! This has been done here four times in the last six months. There's one woman here on a job now for nine months. She applied to have the job and they told her no."

Citing a lack of support from the union and management alike, the women said that challenges from women in similar situations were rare. They saw the policy as a typical exploitive, capitalist, profit-motivated, cost-cutting measure that ultimately benefited the company. They were well aware that male managers and male unionists generally operated in synch to preserve male privilege and power underground. They knew that their subordinated group status resulted from men's solidarity, regardless of race or class, and they responded with varying forms of resistance. Simply being aware of the situation was the first step in resisting. But for other women, it was the basis for overt rebellion. As Darlene put it, "The women don't want to stand up for their rights. They aren't willing to push it." Only a few of the women at the mine, she said, were "fighters" like her because, in her view, "if you want to advance, you got to make waves."

Seniority and the Posting and Bidding Process

For any unionized workforce, seniority is a basic criterion for promotion. But for women in nontraditional jobs, it has been a formidable barrier (Deaux 1983). According to the contract, only miners with seniority, defined as the "length of service and the ability to step in and perform the work of the job at the time the job is awarded," may bid on a vacant position. Job vacancies must be posted "by management in a conspicuous place at all portals of the mine for a period of (5) five calendar days, but no less than (3) three production days," and are to be "properly identified as to portal (underground mines), job title, wage rate, at nonrotating mines as to shift, and the most recent measurement of respirable dust in the work place (underground mines)." Applicants' names are put on a list. Then, "within (3) three production days after the end of the posting period, the senior employee having the ability to perform the work of the job (including panel members) making a bid for such permanent vacancy or new job shall be selected from the applicants." During the posting and selection period, "management shall have the right to fill such vacancy with an employee they may select" (UMWA/BCOA 1988, 97–107). Thus management and, more directly, bosses evaluate miners' ability to step in and perform a bid job.

During my discussions with miners, I learned that their seniority in years was localized; it was based on the time spent working at a mine located in their local union's district. In 1990, data from posted documents in the lamphouse revealed that the most senior miner began working in 1970, while the most recently hired miner began in 1984. Hiring at the mine occurred in two waves that reflected the relative health of the industry. The first wave occurred between 1972 and 1974. The second wave, when most of the women were hired, occurred between 1978 and 1980. The men's seniority ranged from six to twenty years; the women's seniority ranged from nine to fifteen years. Specifically, Sadie, the most senior woman at the mine, began in 1975, while the least senior woman, Michelle, began in 1981. Although their seniority in years was relatively low, the majority of the women were not among the most junior miners. Based on their seniority, women could compete with men for posted job vacancies. "The only men that have time on us already got like eighteen to twenty years there," said Lynn, "but other than that, the first women hired there, they got pretty much time in now. Too much to give it up." As Michelle said flatly, "A lot of them women's got the seniority to bid over half them guys out." But there was little correspondence between the women's seniority in years and their classified job grade. For instance, Sadie spent her entire fifteen years as a beltman (Grade 1), while Michelle had a Grade 3 job. She attributed her advancement to good fortune, saying, "I think a whole lot of people overlooked it [job vacancy], was the reason that I did get it. I started not to bid on it myself because I am at the bottom of the seniority list now. But it was overlooked, I got it, and I was glad." Despite her doubts, Michelle's perseverance had paid off.

However, most other women thought their seniority was sufficiently lower than the men's and they simply didn't bid on more classified jobs. Alice, who had been a GI during her entire mining career, said with a smirk, "You can bid on them, but that won't guarantee you'll get one. Seniority people get them. It's hard to get a job to bid on [with] so many people. People here are so young, by the time they retire, we'll be ready to retire with them. The old buzzards that are here who could retire, won't retire. Sometimes you might be lucky enough to get one that somebody else don't want on a different shift." Previous studies about women in nontraditional occupations have shown that the "perceptions of opportunities are in part dependent on evidence that members of one's own group occupy particular positions within the organization" (Deaux 1983, 292). In 1990, eighteen of the twenty-three women at the mine were classified in Grade 1 jobs. While most of the women at the mine knew that most women were GIs, they were also unable to name those few who had more highly classified jobs. By 1995, little had changed. The women's perceptions of themselves as individuals were based on their perceptions of themselves as a group; since they knew of so few women in higher positions requiring more skills and responsibility, they considered their chances for advancement fairly dim. Lynn, a roof bolter (Grade 5), declared, "There might be three women that's classified. Sadie and [another woman miner] is belts, but them women been on belts ever since they been there. People could've bidded on what they wanted and pretty well got it. . . . There's been buggy jobs, there's been roof bolter jobs and miner helpers jobs. They've had to choose to do that."

As far as Darlene was concerned, the women were "not psychologically up" to advancement, becoming discouraged after an unsuccessful job bid. "At first they would bid on them," she said, "and then when they wouldn't get it, they wouldn't bid no more." Alice, a GI, admitted to bidding on a job only after I questioned her persistently. Turning away from me with her shoulders suddenly drawn, she said hurriedly that she had never bid on a more highly classified job, then added, "I mean, I bid on it, but I couldn't get it. Somebody else with more seniority got it. Like with shuttle car operator. I bid on shuttle car operator, but I couldn't get it. Somebody else had more seniority."

Aside from time on the job, seniority was also contractually defined as being qualified to perform a specific job at the time it was awarded. When miners bid on a posted job, management can evaluate their performance according to "safe operating procedures." Several of the women miners said there were instances in which men with fewer years on the job had been given classified jobs over more senior women because bosses believed the women were not qualified. And even though some men had to prove they were capable, said Carla, "nine times out of ten, even if they didn't like the man, they would give it to him over the woman." While it was possible that a woman was simply not qualified to do the work of a more classified job, sometimes the evaluation process was little more than a thin veil for sex discrimination. As Carla put it, "That's the catch. If you got the seniority [and] you put your name on that paper, that isn't going to do you a bit of good [if] they feel like she can't do this or he can't do this. They'll put you through a test. Oh, they tested Darlene on that wire job when they finally gave it to her."

Darlene told me that before the job was posted, she had been working as a mason near a wireman hanging trolley wire for the mine cars. After getting some on-the-job training by helping him occasionally, she decided to bid on the job. But the company awarded the job to a man with fewer years' seniority, saying she wasn't qualified to do the work. "Hell," she remarked, "I'd been hanging wire for almost two years," and she filed a grievance. Six weeks later the company gave her the job on a two-month trial basis, after which she was awarded the job permanently.

Black women miners' difficulties were compounded by racial bias. Renee, who had a Grade 3 longwall job, attested:

> Me and the same guy bid on the same job, different shift. They let him go on his job, but they never did let me go on my job. I had more time than the guy who bid on the job. I'm like, what's the problem? The union come to me and asked me, "What's going on?" I said, "I don't know." This guy was on his job two weeks, maybe more, and at the same time, I said, "Why can't I go to my job?" I go upstairs to check with the boss over the longwall and he said, "We got to test you." I said, "For what?" "To see if you can lower the shield." I said, "I've done this job before." Then another guy came back and told me [that] the boss don't want you on that job. So I asked the boss on the elevator, "You don't want me on this job?" He said, "No." I said, "Well, some of your crew members came back and told me." I said, "I don't have no problems with you or for you. If there's a problem me working for you, it's your problem." I went upstairs and told the super about it. He said three of his crew members came in and told him. I said, "If I cannot do this job," which I knew I could, "then I want from now on anybody that wants on

> it is going to be tested on it." I ain't saying a word, but the next time they go up, they're going to be tested on it. They took [a male miner] off his job because they had to test me. They knew I was going to bust them wide. While I was waiting to be tested, they took him off his job and made him retest because I had to test on mine. I said, well, ain't that something.

She believed that what happened to her was racially motivated. "I really do think that. But I'm not going to argue. I'm going to do it the right way. I'll just wait and go through procedure, [but] that's discrimination."

Although most of the women and men miners insisted that the company adhered strictly to the seniority rule outlined in the contract, there were other instances where management more blatantly violated the job-bidding policy, demonstrating both gender and racial bias that favored white men. Several women, including Carla, told me the same story. She said:

> There was one lady that bidded on a motor job [Grade 2]. She got the job. They let her run a motor one day. A guy [who] was on vacation and didn't bid [on the same job] came back, and they took her job away. They said he was on vacation and he didn't have a chance to bid, because they wanted him on the motor and not her. That was a Grade 2 [job] and they busted her down to Grade 1 so he could outbid her while he was on vacation. By a general rule, whether you're on vacation or you're on sick leave or you're on compensation, and you don't bid, that's it. No! They stepped in this time and made an exception for this guy so he could bid and get her motor job.

Even though her contractual right had been directly violated, she did not file a grievance over the matter. Rather, she told a woman co-worker that the next time a foreman asked her to temporarily fill a job, she would simply tell the him she was not qualified to do it. Renee and Michelle confirmed racial bias in the awarding of jobs when they told me of an incident involving "a black man [who] bidded on a [highly classified] job. They actually called him in the office and had him behind the door. Terrified him, telling him he had to go behind the jack line and stuff," trying to convince him that the job was too demanding and dangerous. As it turned out, his would-be co-workers, "this girl and one of the white boys on the section, said, 'We didn't want to do his job for him.'" Previously, he had been "a buggy person [a shuttle-car runner], and once he bid on the job, his job was for bid. So when they scared him out of taking the job, he had to go back to GI. I saw this man hanging rag and this man was fifty years old." For women who had held classified jobs, advancement was a matter of resisting the misuse of organizational practices.

One more factor that affected women more than men was the posting of highly classified jobs on preferred shifts, particularly for women with childcare responsibilities. Some women said that staying on a certain shift was more important to them than promotion. For example, Kelly, who began working at the mine in 1979, said:

> There's not as many jobs going up on the board to bid on. Someone has to leave. Say if a roof bolter job comes up on day shift, the only way that happened is if the day-shift roof bolter went somewhere else, which is unlikely [because] he wouldn't be on day shift anyway. The ultimate shift is day shift. The opportunity's not there for anyone. Actually, I could get a roof bolter job on hoot owl or evening shift, [but] I do not want to go on evening shift. I prefer day shift, which is totally out of the question. If the opportunity is on hoot owl or evening shift, they are low seniority to us, basically, but who wants to work on a hoot owl? That's why I'm not a roof bolter or on any other job.

Carolyn, who once had six children at home, agreed that staying on a certain shift outweighed the advantages of bidding on a more highly classified job. "That's mainly it. They don't want to go to another shift," she said. "When I was on hoot owl, I had to have evenings because of the kids. I didn't want the kids alone all evening even though they were teenagers; you don't leave teenagers all that time. So that was my basic reason for not being on evening shift." Commenting on Lynn, who was then a classified roof bolter, Carolyn added, "There's one woman who's worked the hoot owl shift and she likes it. But she doesn't have any children. The oldest one graduates this year and her other kids are gone, so it's probably a better shift for her." Typical of working women's "double burden," the women miners' unpaid labor at home, especially when caring for children, took priority over their paid labor status at work.

In more recent years the women's difficulties getting classified jobs have been compounded by the economic woes besetting the coal industry. Since the mid-1980s, no new miners had been hired at the mine, and miners doubted that there would be any new hires ever again. Since then job vacancies had also seriously declined, and by the mid-1990s they were very rare. During our 1990 interview, Carla, who began working underground in 1980, told me, "They posted jobs every week. I'd say the first six or seven years I was there, you could've bidded on a job if you wanted to promote yourself. It's just got in the last maybe three or four years that they haven't posted a lot of jobs." Miners I spoke with in 1990 estimated that no more than four to six jobs had been posted annually, which they attributed to the company's cost-cutting measures. Several women criticized the company, saying that management would do whatever it could to get by with the personnel it currently employed. "If I was the company," Jessie quipped, "I'd do it, too." But Ellen was not as understanding. She criticized the company for furtively hiring more highly skilled men miners from outside the company to replace those leaving rather than training less skilled senior miners, especially women, from within. The women I interviewed thus demonstrated varying degrees of class consciousness and, more important, varying degrees of awareness of their own subjugation based on the intersection of class and gender.

Realignments

Periodically, as major changes in coal extraction necessitated the relocation of the longwall system or plow, miners were reassigned to different jobs. At the mine, there were realignments once every three or four years during the 1980s, but by the 1990s one or two occurred every year. Before a realignment, miners indicated in writing their job title and shift preferences for management's consideration. They were then reassigned to "available jobs on the basis of seniority and ability to step in and perform the work of the job at the time" (UMWA/BCOA 1988, 99). In other words, because the company could choose to change the number of certain jobs, high-seniority miners were usually retained in their former positions, while low-seniority miners were either shifted horizontally within their former job classification or summarily demoted. This was how classified miners were realigned back to Grade 1 jobs. In 1990, four of the women I'd interviewed had been realigned since their hiring. All of them were demoted from the highest or next-to-highest job classifications to Grade 1 jobs. Carla said, “Every time they've had a realignment, I've been busted back to general inside. Then I have to bid my way back up.” For these miners, realignments were figuratively analogous to climbing a greased pole. Being demoted meant that a realigned miner must begin the bidding process anew if she or he wanted to regain a position held previously.

When I raised the issue of gender bias in realignments, women perceived a definite pattern of discrimination, whereas men saw nothing of the kind. Men reminded me that because some men had the same seniority as the women, or less, both women and men were equally affected by the process. But most women said that men had a slight advantage when realignments occurred, mostly because women had fewer skills than men and therefore fewer options when realigned. “There were guys that would be called back, and you would have, like, five years on them. They would be called back to a certain job because they knew how to do it and you didn't because they never gave you the chance to learn,” said Sadie. “They never gave us the chance to learn how to do it, like rock driller or this or that. Then when you tried to do something about it, well, they were qualified for it. That's how they got by with it.” Realignments also disadvantaged women who had parenting responsibilities. While they may have had enough seniority to get a more highly classified job, they often had to settle for a Grade 1 job on a shift that accommodated their work at home.

Women also cited cases of blatant favoritism, while, again, most men denied it existed at all. For example, a male miner who was also a union official retorted, “During a realignment? To actually just single out a woman? There'd be no way you could do it without it being absolutely obvious to where you'd get burned. I mean, you would just about have to have some kind of special, *special* vendetta against somebody.” He admitted that bias was possible but insisted it was highly unlikely. But, according to most of the women I interviewed, most women and some men were disadvantaged by realignments. Carla and Lynn complained that realignments merely represented a formalized way of showing favoritism. Carla explained, “Say, like, if they put up a bunch of jobs to bid on. People who've got seniority bid on these jobs, [but] they [the company] don't want these people doing these jobs because they got favorites. They want *these* people in these jobs. It's realignment time, *these* people end up with those jobs, you end up with something else.” Carolyn observed that although some men would be demoted after a realignment, "they'd put most of the women back on rag or lowdown jobs." As previously discussed, for both women and men, those jobs carried a stigma because they required no operative skills. Being demoted to a stigmatized job in the hierarchy only exacerbated the problematic nature of women's subsequent attempts to climb the job ladder once again.

Contractually, the company had agreed not to show favoritism or to discriminate when awarding jobs. But the nature and allocation of mine work allowed for a wide range of discretionary relationships between bosses and crew members. Favoritism was thus a central issue in my discussions with the miners, both male and female. Yarrow (1985, 40) aptly points out that favoritism is an "old paternalistic tactic" used to attack miners' unity. Favoritism at the mine, as in other work settings, was partially defined by a worker's orientation to the company, but much more so by the worker's social relations with management, sometimes defined by the most transparent of ulterior motives. Generically, an individual accused of currying favor with the company was labeled a "company suck." But, as I learned through observation and interviews, the label was applied with varying degrees of sincerity. Occasionally miners used the label as a jocular form of address. With somewhat greater intensity, miners envious of a co-worker's status might label him or her a "suck" for working "too hard" and thus violating class solidarity. Kelly, who had a solid reputation for hard work, said:

> With any company or any business, you have your favorites, and it's mostly those people who do anything you ask them to do, work any hours you ask them to work. I'll do basically anything they want me to, but they're paying me to do it, [and] the people that are calling me "suck" won't work. But I take it in stride because I always say—and that's not bad to be called a suck, it sounds terrible—but somebody has to do it and I'm doing a good job at it.

Kelly told me several times that the company paid her bills. Her attitude bears out an observation by Yarrow (1991, 299) that during times of hardship and crisis in the coalfields, "harmony of interests with precarious operators becomes more plausible, and hard work is stressed as a way to keep the company in business and protect jobs." In this sense,

Kelly's loyalty to her union brothers and sisters remained intact and uncompromised.

But the term "company suck," used in its truest and most derogatory sense, described miners who curried favor with management in return for unfair advantages. Sadie described the company suck as "part of the crew, but he's one of their pets. He's a boss's favorite. They call them 'sucks.' Most of them bring the boss their lunch or dinner. Do any little thing they ask them and the boss will get to like them and they get all the easy jobs," or days off upon request, or both. As Yarrow (1991) found, the company suck is a miner who bonds with the boss rather than with his union "brothers" and is "unmanly" because he lacks the courage to stand up to the boss. Some women in the mine I studied had provided sexual favors to achieve the same end and were viewed as traitors to their class and gender. They were particularly disdained and disliked by other women, one of the subjects of Chapter 4. Thus, currying favor directly affected whether a miner was promoted, regardless of her or his work performance and reputation. Several women and men miners said that if a foreman wanted a particular person in a particular position, eventually that individual would get the job, regardless of the formal mechanisms governing promotion and work assignments.

Before the women began working at the mine, interpersonal and kinship-based favoritism resulted in job discrimination against certain men. As with the women, men themselves were subject to patriarchal and capitalist controls. As Ellen, a GI, pointed out, "I'd like to say that because I'm a woman, they keep me where I'm at. I used to think that, but I don't know what it is now. I think it's, in general, they've got you and the company don't want nobody to be nothing but GIs. This company knows they got you. They got the women, and they just might as well like it. The [company] is smart because the more GIs they got, like an opportunity comes if you can run a miner and take that GI. They want you as flank support." However, as Hartmann (1976, 138) has emphasized, patriarchy is "set of social relations . . . in which there are hierarchical relations between men and solidarity among them which enables them to control women." Gaining peer acceptance underground depends heavily on a miner's hard work as blended with masculine-identified traits. As women entered the mines, gaining acceptance among bosses and male miners was highly problematic because of the system of male bonding. As Ellen put it:

> It's the pal system, like if there are two young men, they stick together, and they stick up for the boss, and they'll work their hind end off and you'll find out that the boss has given them an hour of overtime and top pay. I found out if you know a lot about football, baseball, basketball, how to deer hunt, or how to go to a club and drink a lot of booze [*said grinning*], if you was a man [*chuckles*], if you could keep an interesting conversation about a football player, somebody the boss likes, that's what the bosses talk about.

And so, as the women miners adjusted physically to the brute work required of them and began to accrue enough seniority and skills to compete with men for operative jobs, they became more of a threat to masculine solidarity. To some men, they were sex objects to be stigmatized and stereotyped as incompetent workers. Sexual stereotyping completed the process whereby women, because they could not be "run off," were matched with certain jobs defined as "women's work." While men miners and bosses had accepted the women as workers in laboring positions, other stereotypes about the women's incapability for running machinery were in evidence. While all the women developed a dual consciousness, their definitions of self and forms of resistance varied. While some women were determined to pursue more highly skilled jobs, many limited their own aspirations, focused on being competent at lower-skilled jobs, and became invisible by blending into the male-dominated workforce. The following chapter examines how they negotiated their identities as unionists and as women doing "men's jobs."

3

OURS IN SOLIDARITY: WOMEN MINERS AND THE UMWA

When women joined the ranks of men in mining coal, they also joined the miners' labor union, the United Mine Workers of America (UMWA). Historically, Appalachian women have been active in union affairs as the mothers, wives, sisters, and daughters of coal miners. Many were members of UMWA "ladies' auxiliaries," which marched during protests and stood on picket lines during strikes. (For more discussion of their participation, see Giesen 1995; Maggard 1990; Moore 1996a, 1996b; Scott 1995). But unlike most of these women, who acted as supporters of coal-mining men, the women who began mining in the mid-1970s became card-carrying members of the UMWA. By July 1977, there were 858 women in the UMWA's rank and file (*Mountain Life and Work* 1978b).

Initially, the UMWA leadership reacted to the prospect of women members as negatively as the coal operators did to hiring them. At first they paid little attention to the women and their advocacy organizations, such as the Coal Employment Project (Lilly 1989; Thrasher 1981). But slowly women miners nationwide began to make themselves known. Many of them attended the meetings of their locals faithfully; often their attendance records were far superior to many of the men's (Moore 1996a, 1996b). In addition, some women became active in their union locals by serving on mine and safety committees (Dawson 1992; Moore 1996a). Many women miners also walked the picket lines during the 111-day nationwide strike that began in December 1977. Being from coal-mining communities and families with deeply pro-union traditions, these women understood the importance of gaining the union's acceptance through hard work, loyalty, and pride. In this way they could truly belong to a historically strong and often feisty union that would protect their rights in return (Moore 1996a).

Thanks to the success of the class-action suit filed by the CEP, the number of women miners had begun to increase dramatically by the end of 1978. At that time there was a growing recognition among these women and their advocates that women miners were seriously underrepresented among the ranks of local union leadership. The previous year, of the more than eight hundred delegates at the UMWA international convention, only two were women (*Mountain Life and Work* 1978b). It was clear that the UMWA leadership intended to ignore them and their concerns. In an effort to remedy the situation, a group of twenty women miners visited UMWA president Arnold Miller in Washington, D.C., in August 1978. They asked him to issue an official statement supporting women's rights to fair and equal mining employment. Although Miller agreed, he later went back on his word when he refused to ask the International Executive Board (IEB) for a resolution of support. Union officials claimed that the women's request had come too late for them to put such a resolution on the board's agenda for the following month (Moore 1996a).

Over the next few months, in an attempt to pressure the union to pass the resolution, the CEP featured a story in its newsletter that asked, "Which Side Is the UMWA On?" Meanwhile, women miners from coalfields around the nation lobbied their IEB representatives, stressing the need for the UMWA's support against all forms of discrimination in the industry. In response, the IEB unanimously adopted a resolution at its November meeting "that the UMWA support efforts by sisters who are trying to achieve greater opportunities for women in the coal industry as it tries to meet the needs of this nation" (*United Mine Workers' Journal* 1979a, 23–24). Thus, five years after the first women officially entered the mines, women at last began to receive at least the tacit acceptance of their union brothers. Even so, the UMWA, with its 180,000 members, still had no committee for serving the needs of its women workers.

Ironically, the greatest problem women posed to UMWA leaders at that point was their political activism (*Business Week* 1979; Perry 1984). Unable to achieve legitimacy as union members and to have their concerns addressed, women miners involved with the Coal Employment

Project took matters into their own hands. In June 1979 the CEP sponsored the first national conference for women coal miners in Institute, West Virginia. Among other issues, such as mine safety and job rights, the women discussed sexual harassment and the sex discrimination they suffered from company personnel and their union brothers. They also contended that the union's sick-leave provisions failed to give miners, especially pregnant women, enough time off. As part of an overall effort to change the men's attitudes, the women resolved to educate them about women's concerns. Conference leaders encouraged them to participate in union activities and to run for union office. At this first conference, the women also pledged their support for the rank and file and made clear that it was union leaders who were the source of their discontent, despite some pro-female actions taken recently by the IEB.[1] Then, in November, the IEB held its own one-time conference for women miners in Charleston, West Virginia, specifically designed to identify and address the special concerns of women miners and to make them part of the bargaining process in upcoming contract negotiations. Workshops covered women's health and safety issues, contractual rights, arbitration, job bidding, and shift assignments (*United Mine Workers' Journal* 1979b).

In the 1980s and into the early 1990s, the CEP's annual conferences were attended by rank-and-file men miners, local union leaders, and numerous UMWA officials who participated as speakers and workshop leaders. Women miners at the CEP's conferences continued to identify the coal companies as their true adversaries and vowed not to let them use gender differences to divide and conquer the union (*United Mine Workers' Journal* 1980b, 1982c, 1984). They began to increase their union membership and to get elected as officers in their district locals. Despite massive layoffs in the industry in the mid-1980s, these women continued to serve the UMWA. "Some women have been unemployed for months—even years—but a lot have stayed involved with the union, helping with relief committees and running for local office," remarked Bobbi Regan, a woman miner and local union official who attended the 1984 CEP conference (*United Mine Workers' Journal* 1984, 17). That same year, CEP conferees spearheaded the parental leave initiative as part of

1. Earlier that year a CEP mailgram campaign, similar to the one that lobbied for the IEB resolution to support women's right to mine coal, successfully persuaded the UMWA leadership to change its annual convention site from Miami to Denver because the state of Florida had recently failed to ratify the Equal Rights Amendment (Hall 1984; Thrasher 1981).

miners' contract negotiations. Years later, they also participated in non-violent protests during the Pittston coal strike of 1989–90, and served as strike captains and relief coordinators during selective strikes in seven states in 1993 (Moore 1996a). In the 1980s women's representation in the union took a step forward with the election of UMWA president Richard L. Trumka in 1982. Trumka, a lawyer and former miner, had been a charismatic and vocal critic of the UMWA leadership for several years. From the beginning of his presidency, he emphasized the need for UMWA miners to look beyond gender and racial differences to build a better union. During the early years of his presidency, he paid special attention to women miners. At their 1983 conference in Charleston, West Virginia, he told them, "The coal you mine has no gender. It's not female coal or male coal. It's just coal." He said that the women had proved "beyond the shadow of a doubt that a woman's place is in the mine, if that's where she wants to be." A woman miner at the conference remarked that "coming from the president of our union, a statement like that makes a lot of difference to the men. The union is what gives the women and the men a common bond. The important thing is that we are accepted now. I think we're on our way" (*United Mine Workers' Journal* 1983, 14).

In essence, the CEP was an outside pressure group that gave women miners a voice for their concerns and exerted influence on the union (Seitz 1995). Once inside the UMWA, it was incumbent upon them to establish and maintain a balance between their gendered and class-based interests and to create the "space" necessary for defining women's class interests (Fonow 2003). At the international level, women miners had laid the groundwork for an agenda that allowed them to identify as women unionists. However, Milkman (1985) points out that even when unions have addressed women's issues at this level, they have failed to extend their concern to the rank-and-file women in their locals. Locally, male unionists have continued to resist what they perceive to be women's encroachment on their leadership. As in the mine itself, they have "otherized" women by viewing them as special, as in need of protection, and in terms of their traditional female role.

Women unionists, for their part, have often failed to see themselves as full participants in the "brotherhood" of the union. "Unaccustomed to wielding power in the rest of their lives and intimidated from the outset by the prospect of self-consciously maneuvering for power within an organization, women members often decline to enter the competition at all" (Milkman 1985, 306). The alienation that often occurs between union leadership and the rank and file is even more pronounced for women than it is for men. It is not surprising, then, that during their early years in particular, the women at the mine I studied often felt that union officials were unresponsive to their needs and uninterested in their issues. Even so, as hard times came to the coalfields, the women's and men's sense of solidarity had begun to converge, and many of the women had come to realize that one woman's concern was a concern to them all.

THE STATE OF THE UNION

Since the early 1980s, domestic coal companies have sought ways to maintain their profits amid fierce global competition and falling prices by using a combination of strategies—namely, adopting longwall technology, shutting down marginally productive mines, subcontracting production work, and generally keeping labor costs low (Yarrow 1991). Coal operators used the more dire economic forecasts of the 1980s to justify these cost-conscious measures and to reestablish more paternalistic means of controlling the workforce. Since then, the bargaining strength of the UMWA has been in serious decline. Several observers (Perry 1984; Simon 1983) have noted that economic stagnation has had deleterious effects on the coal industry, resulting in layoffs, especially in southern West Virginia and southwestern Virginia. Since women were among the most recent hires, the layoffs have affected them disproportionately. The economic downturn in the coal industry extended into the 1990s and placed the UMWA in a position of relative weakness with respect to the coal operators, encouraging cooperation rather than confrontation (Simon 1983).[2] As one of the local union officials in my study put it,

2. Two major exceptions to the union's relative passivity that changed the relationship between the UMWA and the coal companies were the strikes against the A. T. Massey Coal Group in 1984–85 and the Pittston Coal Company in 1989–90. It was during these two major conflicts, Couto claims, that union "miners reminded capital of the costs of debilitating competition and the benefit of industry-wide standards for wages and working conditions" (Couto 1993, 165). In its unrelenting pursuit of profits, he contends, capital, as represented by the coal companies, has no memory or conscience. It is the miners, with their coal-mining family legacies, who supply capital with both, particularly during strikes.

The Pittston strike in southwestern Virginia was settled earlier in 1990, before I first went into the field. Two of the most memorable protests during the strike were the women's auxiliary's occupation of the company's offices and the men's takeover of the company's

"The militancy is over." Since that downturn began, each successive contract between the two parties has contained "take-back" concessions, leaving miners with fewer benefits and tighter labor controls than they have seen since the 1930s (Simon 1983). According to another union official, "If you saw any of the contract agreements over the years, you can see how diluted that language is, to where no one can understand it, really. Everybody wants to interpret it any way they want. We really went backwards as far as labor. We're just struggling right now to keep our head above water and hope for better times." What a contrast to the situation twenty years earlier, he said.

> We were like big fat hogs [then]. When we ate, we wanted to go somewhere and just go to sleep. The Mine Workers were like that through the 1970s. They lay down somewhere and went to sleep. It's going to be hard to ever recover from it. The upper part of the unions at the international levels have stalled to where they realize they need recovery. The companies can survive the strikes, especially when they're strong. I think that they saw that they need to give some flexibility to corporations. But then I think corporations are kind of like the greedy animal or person who will take it all away. I think that we took some steps backwards and we haven't stop running yet.

At this point, he continued, "What we've needed is strength. That's our only salvation, is to merge with AFL-CIO or the Teamsters, if we can all get together." Typically, in the face of crisis and job loss, although miners have reassessed their militancy and justified their concessions to the coal companies, they have continued to define the companies as their greedy and selfish adversaries (Yarrow 1991). At the same time, the same official said that another reason for the union's political decline was "all the preferential treatment. The men were doing all the work and saw the women get preferential treatment [that] has caused a lot of resentment among the men." Apparently, even in the mid-1990s, women still

---

Moss #3 preparation plant (Wilayto and Cormier 1990). These events received national media attention. They were also believed to mark the beginning of a new era of labor-management relations and renewed UMWA militancy under the leadership of Richard L. Trumka. While I was conducting interviews, management at the mine employed Vance Security, the security company that had earned a reputation for its militaristic tactics and sometimes brutal handling of the Pittston conflict.

had not yet shed the stigma for causing "trouble" and dissension among rank-and-file miners. His remark reflects the splintering of class solidarity by patriarchal notions of women's inferiority, which Yarrow (1991) identifies as the most profound impediment to the class-based interests of miners in the Appalachian coalfields. When male miners' gender consciousness undermines their class consciousness, it erodes the union's political power to the capitalist's advantage.

Like the men, the women at the mine also readily recognized the union's predicament. As Kelly explained, "We're not very powerful anymore, and a lot of it, I think, has to do with the way the contract has been written by the company's lawyer and our lawyers. There are so many loopholes, and one word changes the entire meaning. We are barely hanging on as far as I am concerned." Two contractual "takebacks" over which the women voiced concern were health benefits and the no-strike clause. "We won't have a medical card when we retire," Kelly told me. "It's negotiable every contract if you didn't retire in October of '94. Everything is negotiable contract to contract on your medical card." Even though most miners felt that in order to gain ground the union needed to strike, "we're not even supposed to strike with this contract," Lynn said. "The last two contracts we've been under, it's got a no-strike clause in them. In fact, they could probably fine us if we started the strike, but I believe if they ever did strike up there that the company wouldn't be as bad about pushing them as they have been lately. I believe if they'd ever strike [it would] show them." Two more recent issues that concerned the women were the company's subcontracting of work using nonunion labor and threats to lengthen their shifts from eight to ten hours. Now was the time to act, said Lynn emphatically, because "we done gave up all we can afford. If the union gives up anything more, we ain't going to have nothing. No matter how much [the company] makes, they are always going to want to make another dollar. If we're going to have a union mine, we need to stand up now." Even though the superintendent had dared them to do so, she said, going on strike was very unlikely. "Really, they're job scared. There's no jobs out there. When you go to talking about maybe having a strike," she said, many of the men have said, "'we can't do that.' But in the end? They're going to have to do it anyway. They're going to lose their jobs if they don't stop them now." Kelly concurred, saying, "We need to take a stand. You know it's illegal to strike and they can force you back to work, but that's what needs to take place is a nationwide strike." In contrast to the attitudes of some of the men, the women's position echoed the working-class militancy of an earlier era. It also signified their adoption of the more masculine-identified beliefs of union "brotherhood," which demonstrated their assimilation of the class-based interests of miners.

According to the mine superintendent, labor-management relations at the mine had been relatively amicable in the early 1990s. Even though there had been a slight "resurgence of the 1970s belligerence," he said, the union's renewed organizing campaign was a sign that they were now operating from a position of relative political powerlessness. Then newly elected UMWA president Cecil Roberts, he said, was stumping again because he knew the union was dying. In the superintendent's estimation, over the past two decades, the UMWA had "lost its sense of purpose toward fairness, safety, and honest work." As some miners themselves believed, he thought that most miners today were "selfish and greedy" and that their only real interest was in making money. Many of them competed for overtime, while so many other miners in the coalfields were laid off. In other words, their masculine-identified union solidarity was beginning to unravel (Yarrow 1991). The superintendent also said that these more self-serving men were the same ones who had once harassed the women when they began working underground. In essence, class solidarity was splintering because of men's competitiveness and their sense of superiority over women. He predicted that at some point job sharing would become necessary. Given that the women were more "rational" and "more willing to sacrifice," he felt they would be more cooperative than the men. He believed the women did not have the macho "I'll-rot-in-the-pits-of-hell-before-I-give-in" mentality. Because they lacked this "pride factor," he felt the women were more "trusting of the corporate side" than the men.

Nonetheless, he said that much of the women's information about management-labor relations was filtered to them through their union brothers. Based on my information, the women were no more or less "militant" than the men, and they most definitely had their own thoughts about labor-management issues. The superintendent underestimated the degree to which the women at the mine had aligned their class-based interests with those of their union brethren. He was aware, though, that some miners' patriarchal notion about "their" union was a weakness all union miners would have to overcome to fortify and strengthen the union.

Tough economic times coupled with a less powerful union reminded miners that when their jobs were threatened they could fear reprisals for "standing up to the bosses." A typical "survival strategy" among miners is to take a more harmonious and cooperative view of labor-management relations (Yarrow 1991). Although, to varying degrees, the miners with whom I had contact continued to view the company as their adversary, most of the women and some men did demonstrate a conciliatory attitude toward the company. Kelly admitted, "I take pride in my company. I know how I am and I am just a little bit 'company,' because they pay my bills." These miners considered themselves fortunate. The company had laid them off only twice during the past twenty years, for only several weeks each time, after which most of them returned to work. They recognized that the company had its troubles, too, and that in fact it wanted to get out of the coal-mining business altogether. It was obviously in the miners' interest that this not happen. "We are competing to try and stay in operation for [the company]," said one of the men. "Whether I want to be a part of that or not, I am forced to be." As Kelly put it, "If the union and the company does not hold hands, we're both going down. Both of us. We've got to work together. We've got to see each other's point of view. If we don't make money, [the company] will shut the mine down. If [the company] shuts the mine down, we don't have a job.

If you don't work together, we're going down." In adjusting their thinking about the company, the miners emphasized hard work as a way out of economic difficulties (Yarrow 1991), a common theme among women and men miners alike. They regarded "lazy" miners with the utmost disdain and voiced a common complaint about union policies that protected unproductive workers. "What bothers me the most," said a union official, "is the inability of management to be able to manage the workers to where they are productive." After discussing labor-management issues with miners who worked for other companies, he said, he had concluded that the company was "probably the best company to work for. They spoil the people here, really. During the seventies, there wasn't any people. So they took anybody they could get to work. Once they took those people on board with the United Mine Workers being in here, it's not easy for them to get rid of them without a big problem." Other coal companies "would fire half of them," he said. Indeed, several of the miners praised the superintendent for his attempts to keep the company afloat. "I like [the superintendent]," said Kelly. "He is very educated. He is a very good superintendent. He has our best interest at heart as far as keeping that mine running at whatever cost to him, to [the company], not to the union." But, she added, "I don't agree with everything he does because I look at it through union eyes."

Like miners elsewhere in the Appalachian coalfields, the miners I interviewed expressed a generalized distrust of the more powerful elites in "big business" who controlled the company, and tended to blame them for the situation. "They actually control everything," said one man. "Really! They might not actually be down here controlling what you're doing, but they start turning off switches at the top, it'll soon get to you." Thus, it was easier to express sympathy for management, which had to navigate troubled waters with the greedy corporate elite on one side and the workers on the other. "I think at one point there was a management [here] that [was] going to try to do somewhat the right thing," said the same miner. "But now I think the competitiveness is like everything else; [now] they say, hang you." In his estimation, there was only one solution.

> It's really just going to be downhill for working people, because common people don't have the ability to go out and talk them out or produce more lawyers than they do. If I could tell the membership to do one thing that would change their life more than anything for the better, [it] would be to register and vote. Not to go out there and strike or throw Rich Trumka or Cecil Roberts out. The politicians make those rules, they make the rules to where we can't strike. They appoint federal judges for life, they can put you in jail, and they don't have to answer for nothing.

At the same time, however, women and men alike expressed their anger and disillusionment over being "sold out" by union leadership at both the international and local levels. Some miners expressed disappointment in the international leadership's selective-strike strategy for dealing with the coal companies. As one of the men said:

> What killed the mine workers was the selective strike. When Trumka brought in the selective strike, what it was meant to do was to keep the coal rolling so the economy would stay strong. He would let five or six coal mines continue to work and everybody else would be on strike. You'd strike half of them at a time. And as long as the coal was flowing we didn't have no strength

> at the bargaining table. The only way that you can have any strength at the bargaining table is stop all flow of coal, union and nonunion. The law and the courts have got you tied up so tight that you can't legally strike.

Kelly agreed:

> I think the selective strike that he came up with destroyed us. Big companies want to scratch each other's back. If you and I are the coal companies and they choose you to strike, your company, I'm just going to come to you and say, "Look, I'm going to help you fill your orders until this is over." So that's what takes place. I make money and you don't lose your contracts because I'm filling it for you. Then when the strike's over you can have yours back and I made my money off of you. The selective strike doesn't work because coal is still flooding the market. You've got to stop the coal going to the market to do anything. The selective strike does not work.

Moreover, she said, local leaders were not as much at fault for the rank and file's political powerlessness as were those who ran the union, since local leaders' "hands are tied because of the arbitrations. Trumka won't back us. There's things going on right now that he needs to be down here or someone with more power than the local president to try to get contractors in to do union jobs. It's in a grievance now. It's going to stick, too. We need to take a stand. It's illegal to strike and they can force you back to work. But that's what needs to take place is a nationwide strike. [But] it's not going to happen."

Like Kelly, other miners claimed that incumbent UMWA president Trumka, a lawyer and businessman, was too much like the elites who exploited them. In traditional UMWA style, they said they needed someone who could withstand elites' influence and avoid the temptation to "make a deal" at the miners' expense. In short, they needed someone who would really stand up for them, and Trumka had fallen short of the mark. Echoing the sentiments of other miners, another miner who was also a union official said:

> I think people would wish that Richard Trumka would get out. I think that people who are coal miners relate more to a guy like

> Cecil Roberts. Cecil Roberts is the guy who can get out and talk to coal miners. Richard already went over. He was able to take some steps backwards but probably still able to maintain us. We are going to try and get through the largest change we've ever made, stuff we have in our contract. You can't force people to change. The biggest issue right now for us is at the next level. And [during the next contract negotiations] they're probably just going to want to come down and . . . [*claps his hands together*].

According to miners of both sexes in my sample, Trumka had failed to resist either personal or business gain. "I think he's in with the companies," said Irene. "I think he's made deals with the companies of what he can do and what he can't do. I believe that's the way he sold us out. I think he got in there and he's got his bank full." Similarly, Kelly said:

> I don't know Mr. Trumka personally, but I know what I've heard and I don't think he has done us any good. We're paying 2.5 percent of our wages in a strike fund. I probably paid twice as many dues, around $1,500 a year at least, to the union and I'll never see a penny of it. No one will. No one knows where the money is on our level. [I think] some of it goes to miners when they're striking. I think they're using it as their own personal bank to make loans to make money off of money. I don't think it's used in the coal industry.

Kelly also said she thought Trumka had married into a family that owned a large coal company that was "nonunion and as dirty as they come. That's a big issue, too. How can you trust a man who's sleeping with the enemy?"

The miners also expressed a similar general distrust of local leadership, no doubt fueled by the company's co-optation of their former union president during the mid-1980s. When I was in the field in 1990, he was the company personnel officer responsible for settling labor disputes and meting out reprimands. But, as Kelly said, he "used to be our union president. He was president right before the contract strike and he would've been a fantastic president. That's why the company wanted him. [Now] he's totally company, through and through." As could be expected, because "there's always a little bit of hostility between union and company anyway," she said, there were "some real harsh feelings afterwards, like he sold us out." Both women and men miners said they were watchful and had cause to be suspicious of their union leadership lest it happen again. They said they had little doubt that current leaders would be unable to resist the lure of money and an easier job that the company could offer. This was why current union officials had often been unwilling to "rock the boat" on behalf of the rank and file—because someday the company would "buy them off." Lynn explained:

> I think the company offers them a better deal. Lots of times when your union officials first start in, they're really for the union. But after they've been there for a while, they kind of side with the company and they end up getting better jobs. I know that it's offered to them. If I was company, I'd probably do the same thing. You do me a favor and I'll do you a favor. At times they're thinking of how easy it's going to make it on them and they tend to slide a little bit, until some of the [rank-and-file miners] get really radical with them over that and brought it to their attention. And then they'll kind of straighten up and fly right for a while.

There was also a gendered basis for the fraternization between local union officials and mine management. For example, Sally told me, "The district union reps would go to the ballgames with the bosses and the super. They was just like buddy-buddy. The company is giving them good jobs, taking them to ballgames. They're eating steaks." Women miners saw such "male bonding" as posing "an impediment to class solidarity by encouraging miners to see a bond with male managers" similar to the one that emerged when women began working underground (Yarrow 1991, 303). But both women and men said they knew which local officials were reliable and which were not. While all of the miners expressed some level of dismay over the state of their union, they still pledged their loyalty to the local officials who "stood up to the company." Their general sentiment was that, although the company provided a job, the union helped them keep it. This rationale minimized the contradiction between the union's erstwhile militancy and the more accommodating stance it had taken toward the company; it allowed for a softer version of the earlier toughness while it maintained the more conciliatory approach it had adopted during hard times. But the miners' continued distrust of the company and their frustration with management's traditional paternalistic control has remained intact, and their insistence on a nationwide strike has been key to their sense of solidarity (Laslett 1996). They have also recognized that tougher bargaining tactics will be needed during negotiations if they are to maintain health and retirement programs and keep out nonunion subcontracting. In defense of the union, these miners pointed out that there were definite advantages to belonging to a union. As Kelly said:

> I think you do have a certain degree of protection as far as our job stability. Just because they don't like me doesn't me they can lay me off or fire me because of my seniority. If you were nonunion, they would tell you, "My way or the highway." That's how it is at nonunion. If you miss work, they'll just tell you, "Don't come back tomorrow"—and you can't. There's good in the union. Without the union, the nonunion [miners] would be in deep dung because our union scale dictates what they make. So they need us. The nonunion needs us. If everybody goes nonunion, we'll all be working for fifty or sixty dollars a day.

Lynn added her own thoughts on the subject:

> The pay and everything is good, going by the scale. The job safety is good in the union scale. Because they're strictly for safety—I mean, anything that's unsafe, all you got to do is go to the union. They'll get a safety man in there or an inspector. If you're put in a position to where they try to make you do something that's unsafe, if you feel it's unsafe, you just contact your union and they'll get it straightened out. The union will go to the inspectors anytime, especially state inspectors, anytime the company will try to make you do something that's not safe. [At nonunion mines] they can call a federal inspector or something like that about something, but they usually don't because they don't have anybody to back them. See, the union does back you in times like that.

Furthermore, the women expressed pride in what the union represented despite the shortcomings of local leaders. "I do believe in the union," said Irene, now retired. "That union up there has went down a whole lot. They sold us out. They really have. [A former union official], he'll talk real good to you, but he goes behind your back. I go by that book [*pointing to the contract book on the table in front of us*]. I'm a union member. The union put you in the mines to work. That's to hold your job. That's to hold your seniority. That's to hold your head up."

WOMEN AND MEN IN THE UNION

As in other coalfields over the past twenty years, the miners in my study had become discouraged and exasperated with the weakening of their union and many had withdrawn from actively participating in it. Union leadership was chosen by local union members and meetings were held each month, but attendance at these meetings was routinely sparse. Unless there was an extremely pressing issue, said one local union official, "nobody participates, hardly. We've tried over a period of time that I've been here [to increase attendance]. We've gave away numerous door prizes trying to get people to come in. Turkeys and jackets."

One Sunday afternoon in November 1990, I attempted to sit in on a union meeting, but the union president denied me access. Under normal circumstances, I could have attended, he said, but there was important business to discuss. Carla told me later that members of the local had discussed the company's proposal to extend shifts from eight to ten hours, although the current contract held the company to eight hours.[3] Standing outside the doublewide trailer that served as their union hall, I counted approximately thirty members of the district's rank and file in attendance that afternoon. During our interview, Kelly commented with dismay, "We have some five hundred union members and you might get twenty at a union meeting. Yet we want [the current president] to do everything for us." Likewise, a former union official told me:

> The only time they come to the union is when they get into trouble. If they miss a lot of work or they have problems, then

3. The company eventually instituted a shift-change policy known as "hot seat change out." Kelly described it to me: "What we do now is . . . day shift starts at eight o'clock. Evening shift starts at four. And the third shift's at twelve. Neither shift can leave until the shift after them gets there. They can keep you up to two hours. But normally it's only a nine-hour shift. That is part of the contract. They can have alternate work schedules, but the union has to agree on them. They would like to have a ten-hour day, [and] a forty-hour work week means no overtime until the forty hours are over. You really couldn't make much money like that."

> they come to a union meeting. People seem content with drawing their money, their work assignments. They're content, they're happy. But when they're at work, they cry and complain about the union representatives drawing too much money off the union, how the union executives are robbing the men. But come union election time, they don't do nothing about it. Coal miners just like to cry, I guess.

Both women and men miners commented that "the union ain't what it used to be." They recognized that "if the union's not strong it's our fault," but they said nothing about what they would be willing to do to change the situation. And while they praised the union as their sole and final means of support, they also did little to improve it. As noted earlier, the misalignment of the rank and file's class interests with the union's actions had led to the miners' increasing dependence on the company. Given the concessions of the previous two decades and the extent to which national and local leaders had sold the miners out, these miners demonstrated a growing sense of relative powerlessness conveyed through their nonparticipation at the local level, a phenomenon witnessed by other researchers as well (Gaventa 1980). This nonparticipation also reflected the changes in Appalachian miners' class consciousness discussed by Yarrow (1991). Relative to their collective sense of class solidarity in the 1970s, over the following two decades miners became increasingly concerned with their own individual interests and their families' well-being, demonstrating a general decline in their sense of class interests. The danger of this shift was evident in a warning by a local union official, who said, "People don't know what they have until they're going to lose it." Similarly, Carolyn commented, "We're not too strong, and not enough of us attends like we ought to, but we should. It's sad. We're lucky to have what we've got."

Generally speaking, both women's and men's attendance at union meetings was dismal. Since meetings had usually been held on Sundays, the women told me that they were often unable to attend because they ended up working that day. Others said that, like the men, they preferred to spend time with their families because Sunday was their only day off. "There's too much arguing going on," said Carolyn. "You've got one day off. Sunday. I'm not going to spend that time listening to that kind of stuff. I already sacrifice [union dues], and we give them our opinions about what we want, but they go do something else.

So I think it takes too much to get involved." Kelly said with some disgust, "The first [meeting] I went to turned me off totally because it's fussing and cussing and swooping around, the same gripes. Nothing ever gets accomplished. That's the biggest downfall of our union, is the lack of interest. We're hurting ourselves."

But those few women who had continued to attend regularly disagreed. Irene said of the other women, "A lot of them said they didn't want to go because they argue all the time, but that's the name of the game! That's what the union's for! You have to argue your troubles out with it. I didn't see it that way, and if everybody did, we wouldn't have the union."

While at least a few women had been as active in the affairs of their local as the men were, only one woman had ever held a union office, briefly during the 1980s. A former union official told me, "She got involved with the union and was really trying to represent the women. Then she got involved with the international guys, too. She forgot the local level. What she started out to do on the local level was good, but the more she got involved, the higher up she went. She went all the way to the international. She'd get leave from the local and spent enough time up there. They just covered her up. Blowed her plumb out of the water." Eventually, he said, this woman left the union, quit mining, and changed occupations. This woman was unknown to the women I interviewed. Although they felt that women needed to assume positions of leadership, they considered this too difficult. Ellen, who was relatively active in union affairs, said she wouldn't run for office because she was "scared they'll frame me up" or that she might get in over her head. She hoped one day to become a member of the mine safety committee, however. Other women who had children said their responsibilities at home made it too difficult for them to become more involved in union affairs.

Moreover, given that most women at the mine began working there in the late 1970s, union leaders had only recently taken up "women's issues." During the same union meeting, the local elected to send Carla and Jolene to an upcoming AFL-CIO meeting in Charleston, West Virginia, on that very topic. In 1996, a union official told me, "We've been tiresome as far as trying to get them to participate. We had Ellen participating a little bit. She'll be a delegate going to Charleston at the end of this month. We did send Ellen and Bridget to Alabama to the AFL-CIO women's conference." But while the local had continued to send interested women to different labor conferences over the years, he believed that they were really "using the union at times to get out of work. They get time off from time to time to go to attend seminars and we pay for it." Echoing this sentiment, a former union official added that "the women don't usually show up [at union meetings] until they have a women's conference somewhere out of state or in state that a couple of them want to go to get out of work. If they want to go party for a couple of days, then the union pays them to go. The company gives them leave to go and then after it's over they quit coming to the meetings." The implication was that women were less politically committed than men. But just as some men had doubted the women's ability to work in the mine, they also doubted the women's dedication to the union and the class-based interests it represented. In writing about women steelworkers, Fonow (2003) notes the "masculine" consciousness of male unionists and the extent to which women challenge the cultural meaning of class solidarity. Women's challenges provoked a defensive reaction in men, who differentiated women symbolically and relegated them to the margins of the "brotherhood." The men at the mine thus questioned the strength of the women's class consciousness and downplayed their activism, making women's attempts to negotiate their own identities as unionists and to achieve solidarity more difficult. Unfortunately, the men's myopic approach limits the union's strength and overall effectiveness in resisting capitalists' demands.

The women at the mine felt that their reasons for going (or not going) to union meetings and conferences were no more self-serving than some of the men's. Because union officials were also still employed as underground miners, they were granted time off to conduct union business. Lynn said with some exasperation, "Lots of days they don't even go underground, but they are paid by the union for doing union business. A lot of days they don't even have to go in and perform their job." Even more damning, as Ellen complained to me, some men used the international meetings to take time off work. "Did you hear about the big Miami trip in September, where they're sending seventeen members, $25,000 or something?" she asked. "See, they're getting their two weeks' vacation, and then they get two weeks at Miami Beach. I would love to film some of that because this one guy said, 'Are you taking your wife?' And he said, 'Hell no. I plan on finding somebody down there.' Now, that's exactly what he said." Despite the men's view of them, these women had developed their own self-image as unionists who were every bit as sincere about pursuing their class-based interests as their union "brothers," if not moreso. In interview after interview they pledged their allegiance to the union. A few of them also became involved with the Coal Employment Project during its final years of operation. In fact, I ran into Ellen and Renee at the CEP conference in 1998 and again in 1999. They were committed to staying active in the union so they could be informed about issues affecting miners of both sexes. As Renee said:

> I'm a very vocal person at the union meetings, because you work with them and try to get involved to keep the whole thing going. When you're going, you learn a lot of the tricks, a lot of things that could happen either in your job or on the job. I attended the women's conference. I learned a lot there as far as union representation. I went to one in Welch not too long ago and they taught me [about our] rights and [they] teach the women how to know harassment and procedures about that, and it was really great. When you get involved in it, it's like, "How can I help with this problem?" You're looking on the job and you're looking at work [to see] if this is really happening and how we can do strength together. You got to bleed to get anything. I told [the union,] "Anything you want me to do for you, let me know."

Those women who were most active in the union also demonstrated a greater feminist consciousness that had been honed through their participation in outside advocacy groups, such as the CEP. Being subordinated by both patriarchy and capitalism, their marginal status afforded them a privileged view of how both forces operated (Seitz 1995). Often they attempted to insert a feminist agenda into their local's agenda. Their task was to show how women's concerns were also men's concerns, thus thwarting the gendered nature of all miners' class consciousness and activism. By minimizing patriarchy's influence, they focused on developing a more heightened collective class consciousness and more active resistance to the company. In sum, most of the women at the mine were determined to work with men unionists—those who had always been receptive to them, those whose stereotypical attitudes have softened toward them, and those, a dwindling number, who were still opposed to their working in the mines at all. Like Trumka, when he said that coal was neither male nor female, they struggled to demonstrate that unionism was neither masculine nor feminine. Most were ready, willing, and able to join in the spirit of solidarity and do their part to build collective resistance against the forces of capitalist exploitation. In principle, they had a basic belief in what the union represented. Their only wish was that the union represent and support them in return.

UNION SUPPORT FOR WOMEN MINERS

The women's feelings about the union and its support for them varied widely. Some praised their local union officials and said the union supported miners of both sexes. Others said that men unionists' support was minimal and sporadic for miners in general. Ellen allowed that "if I had some [union] people backing me . . . but I'm by myself. That's how I feel and everybody up there feels that way. I think being union should be all for one and one for all. But I think union is union assessment, union dues, and that's all." A few women, however, believed that there was even less support, and sometimes none at all, for them in particular, depending on the issue involved. Michelle declared, "Our union ain't worth a shit," and she went on to criticize local officials for failing to support women in just about any situation. An incident in 1990 certainly contributed to their feelings of alienation. Carla, Darlene, and Rachel told me disgustedly that the union president had held a meeting for all miners in the men's bathhouse. Of course, all but one of them were reluctant to go in. "We're married women," Carla snorted, "we can't go in there!" Only Darlene waded into the meeting, saying, "It's my union, too, you know." On her way in, she surprised one of the men who was hurrying to put on his pants and stammering an apology for being undressed. Quick on the uptake, Darlene answered him, "Aw, it wasn't any big thing." As the story went, the double meaning of her remark elicited gales of laughter from her all-male audience. Her humor successfully eased the tension her presence had created, and she remained the only woman in attendance. Meanwhile, the rest of the women were content to sit and listen through the wall. Later they complained to the union president about the location of monthly meetings, but he flatly refused to hold them anywhere else. "Why couldn't he have it in the lamphouse and shut the door?" Carla said. "His excuse to us was, it was more convenient to have it in the men's bathhouse." The message was clear. They could listen to the men but not be seen or heard, and therefore could neither fully nor actively participate in union affairs. They were offended by the obvious discrimination against them and felt it diminished them as unionists.

Irene had a much more accommodating and complacent view of this episode. "One time they was having a union meeting in the men's bathhouse," she said matter-of-factly. "Some of them said, 'Wait a minute, we can't go in there.' They asked, 'Why can't you bring it out in the open?' [The answer was] because they didn't want the company to hear it all. It's just behind the wall. They can hear it anyway. We sat beside the wall in the men's bathhouse so we could hear it." Unlike some of the other women, Irene's strong union identification outweighed her feelings about the men's subordination of women. A few years later, a new union local president was elected, and all meetings were held off company property in the trailer purchased by the union.

When women did complain about union officials' lackluster support, their complaints concerned sexual harassment, on-the-job training opportunities, or grievances they had filed. In the early to mid-1980s, women miners at other mines in the Appalachian coalfields had been filing lawsuits against their companies over sexual harassment and being denied advancement.[4] The company issued a rule against any

4. Regarding sexual harassment, in June 1981 eight women miners filed suit against Consolidation Coal Company's Shoemaker Mine in Wheeling, West Virginia. The women sought $5.5 million in damages, alleging that they were sexually assaulted, that they were constantly exposed to graffiti about them in the mine, and that there was a peephole into the women's shower room. Rather than file a union grievance, the women decided to prove their point by going after monetary damages (*New York Times* 1982). Although it was never confirmed, several women at the mine I studied said it was rumored there were peepholes in their bathhouse. Kelly told me, "There have been instances where supposedly we had holes in our bathhouse wall, okay? I think, yes, we did. So they corrected the situation and now they have a brick wall, a steel center, and another brick wall." Jolene said, "We had our usual holes in the bathhouse. I came in one day. I looked over at the wall and I could see sunlight. I go over and look through the hole and you can see in the shot fireman's office and into the yard down there. I got real upset. They sealed a plate over it, [but] there's still a doubt in the back of my mind that there's probably a new one over there. I feel like there probably is."

Regarding advancement, in the summer of 1989 the West Virginia state supreme court issued its final decision in a discrimination suit filed against a coal company by a woman miner, *Bishop Coal Company vs. Brenda Salyers*. Salyers, aided by the West Virginia Human Rights Commission, complained that her employer refused to promote her to a scoop-operator job. The basis of her complaint was that she was required to take a test to see if she could adequately operate the scoop, but that none of the men at the mine had ever been tested before being assigned the same job. She was awarded $400 in back pay, the difference between the wage rates of the job she held and the job she bid on, and $2,500 for emotional distress (Fleischauer 1989a).

form of harassment underground and told women to report infractions directly to the company, which in effect usurped the union's authority. When a boss was involved, the issue was between the company and the union. But reporting another miner to management violated the union oath of solidarity. On the one hand, the superintendent complained about the union's inability to deal with the social and sexual conflicts between miners, saying that resolving these issues was like "dealing with children." He deemed it necessary to resort to the same paternalistic control typical of that reasserted during the 1980s by many coal operators in the Appalachian coalfields (Yarrow 1991). On the other hand, a union official said, "The women could have come to us." But most of the women felt that nothing would be done. So they either dealt with these situations themselves or threatened their harassers with the company rule, both of which strategies were generally effective. Similarly, while individual union brothers might show some women miners how to run machinery, the union never initiated an official training policy of its own. Rather, the "training rule," as the women called it, was the result of company policy and not pressure from the union.

Another major area of concern was union officials' lack of support for women during the grievance process. A few women felt that the union did a good job of supporting them, but several others complained bitterly. Generally speaking, those who complained had filed grievances over gender-related issues, while those who praised the union had little or no such experience. For example, Ellen learned from a man on her crew that for several weeks all the male crew members had been given "top pay," while she had been paid a lower wage for the same work. She filed a grievance and eventually received the difference, but she considered this a hollow victory because the union never pressured the company to admit to any wrongdoing.

> All I wanted was what was owed to me, not to set me up and give me top pay maybe for three weeks and then: "Well, we paid her top pay a little bit to get her off our hind end." It didn't do me no good to stand up, I mean, I just lost. The grievance was discrimination. I wanted to know why they didn't give it to me when they was giving it to the men. It made me so mad, and what made me worser, the men thought that they could pay me off. It wasn't the money part, it was the way they done me. I just wanted to show them that they wasn't going push me, and

> they wasn't going to get something over on me, which they did. They won, I feel.

She concluded, "If there was a grievance filed, the president of our union right now and [a company official] is just like brothers. I mean that's just how it is. Heck, it's just messed up bad. They [union officials] get it for theirselves and the hell with you." At stake here, in Ellen's view, was the principle of gender equity and male bonding taking priority over union solidarity. She simply wanted to be treated like other union miners, regardless of gender. The way the matter was resolved left the solidarity among men miners intact and left her feeling "paid off" for accepting her own marginalized status.

According to several of the women I interviewed, the company had discriminated against two women over job assignments, adversely affecting their advancement. In both cases, described in the last chapter, the union failed to back them. In the first case, Dawn had been temporarily assigned as a fireboss for nine months and had requested that the job be posted per the union contract. The company had refused, which meant that she was unable to bid on the job as a permanent assignment. Michelle remarked, "If we had a good union, they could've kicked on it. They've done that for the men, they could have done it for her, too. She should have been one of the women [fire]bosses here, probably would have been the first." In the second instance, Jessie was required to train three male GIs so they could successfully bid on a job for which she had more seniority in both years and qualifications. When she attempted to file a grievance, union officials talked her out of filing. But they later backed a man who filed the same grievance over a different job. She concluded that "the union people got good jobs and they don't want to mess that up."

Both stories demonstrate the strength of patriarchy over class solidarity among unionized miners. On other occasions, some of the women simply decided not to seek redress because they believed that the union would not support them. Carla said, "I just think that some of the union officials that's in office doesn't approve of us women being in there. And they won't help us, they won't stand up for us like they would a man. It's just obvious. If we have to go in the office, we go get one of them." Then, with a hard sigh of disgust, she said, "They sit there. They don't open their mouth, they don't try and defend us or nothing. Nothing! I've had it happen to me."

However, a former union official declared that the union backed the women "100 percent. We give them as much as we give a man." He went on to say that women and men miners alike make it difficult for the union to support them, particularly those with poor work records. For example, he said, "A couple of females one time left the work area. They was suspended for three days. They got mad at me because they felt I should have protected them. When you are scheduled to work from 12:00 to 8:00, and you sneak off and leave the mine at 6:00 A.M. in the morning before the shift is over with and the foremen is looking all over for you, he can't account for you, and you're gone. The union can't help you." He also told me:

> When you fill out an application to come in the mines, they ask if you've had any injuries, any leg breaks. A lot of the women failed to put on there that they had previous car accidents. This one in particular had a car accident and suffered a lower back injury. She failed to put it on her application. Well, she went out on compensation hurt. The company started checking her previous record and found out that she had been in the hospital and they discharged her for falsifying her pre-employment record. She blamed the union, [saying,] "The union didn't stand up for me." What can you do?

Coal companies have always taken a hard-nosed stance toward miners' grievances (Simon 1983). One union official said, "This grievance procedure stuff [is] forced on people. It's a sign of weakness on the UMWA's part when they can force you to accept a procedure that you know you can't win against." The union officials I interviewed stated that since so many precedents had already been set through arbitration, union officials would often advise women not to file. Not enough of the women were aware of this, they said, which led to the perception that the union failed to support them because they were women. According to the women, this was only partially true. Although they admitted that some women were perhaps unaware of how certain precedents affected their grievances, most of them understood that such precedents prevented any miner from successfully filing certain grievances. Ellen said, "There's so much past practice stuff, there's not much you can file that it ain't been done here or been done there. So they don't push it real hard. I think that's part of the problem with some of the grievances." As

Kelly explained more fully, "You have an arbitrator that decides everything for you, and 90 percent of all grievances that go to arbitration, the company wins. Once a grievance is arbitrated, that stands forever for everybody. Not just [the company but for] everybody in the coalfields. Everybody. So you know you really have to be selective when you file a grievance and take it to arbitration." Otherwise, "If you don't know what the procedures are, you can't just file every little grievance you get ticked off about, because all they do is put you up on a little computer screen and say, 'This has already been filed and we lost it.'" As far as she was concerned, a miner should file a grievance only when there had been a serious breach of his or her rights. "At first something pisses you off and you blow off steam in the bathhouse. We've filed a million grievances," she paused and then said dryly, "in the showers." Then you "cool off and reconsider." So, to some extent, there was evidence to support the women's claim that the union supported men more than it did women. The women's awareness of this was often collectively constructed during their daily discussions and debates in the "free space" or "safe space" of the women's bathhouse.[5] This was where they collectively examined the contradictions between their own self-definitions as union miners and the images perpetuated by men who devalued them as "other." Although women often individually resisted men's images of them as physically weak and politically powerless, these "gripe sessions" were empowering because they increased the women's collective awareness of gendered issues and built solidarity.

In sum, the women at the mine were marginalized as miners and as unionists, but where the union was concerned, they were disadvantaged less by what was done to them than by what was not done for them. According to several of the women, even when certain union-backed

5. Couto (1993) regards strikes as "free spaces" beyond those normally found within miners' families and their union. As originally defined by Evans and Boyte (1986, 17), a free space consists of "public places in the community . . . the environments in which people are able to learn a new self-respect, a deeper and more assertive group identity, public skills, and values of cooperation and civic virtue . . . settings between private lives and large-scale institutions where ordinary citizens can act with dignity, independence, and vision." In the context of African American women's quest for self-definition and voice, Collins (1990) refers to these spaces as "safe spaces" where safe discourse occurs without the oppressiveness of hegemonic ideologies and objectification by members of the dominant group. Thus, free (or safe) spaces are places over which a group has control because they are relatively removed from dominant-group members' control. Members of the subordinate group can share information, expand their own individual identities, and construct a network that fosters their potential for collective resistance.

provisions operated in their favor, those union representatives charged with their even-handed implementation had, at crucial times, failed to be effective when a woman was involved. Of interest here is how these experiences have affected women miner's beliefs about their union. Overall, the women I interviewed demonstrated varying levels of union loyalty and varying levels of awareness of how the "brotherhood" let them down. Only a few cited these situations as the reason why they no longer believed in the union and therefore would never participate in its activities again. Some were also quick to point out that there were times when men had suffered the same abuses because of the union's weakened position with the company. Most understood that their failure to participate in union affairs meant they were missing an important opportunity to assert their class consciousness, solidarity, and resistance both as workers and as women, regardless of the negative experiences they may have had.

The women's situation lies at the intersection between gender and class. Women's status as both workers and unionists simultaneously challenged male privilege and working-class solidarity because class solidarity among miners has traditionally been identified as masculine (Yarrow 1985, 1991). Historically, according to Fonow (2003, 86), trade unions have experienced a tension between "balancing the need for worker unity and class solidarity with the need to respond sufficiently to the specific concerns and the unique needs of different groups of workers." Union officials fear that acknowledging these differences will jeopardize class solidarity, and this fear makes them reluctant to support charges of discrimination by minority members, whether they are against the company or the union. As Fonow found with the steelworkers' union, UMWA officials have tended to mute the interests and concerns of their more marginalized workers. Her findings coincide with Yarrow's (1991). As women have entered their domain, men have continued to control the resources commonly associated with full-fledged membership in a male-dominated unionized work force, and they have been less than willing to share them with women because they value their male privilege over class solidarity. Thus, the greater union support enjoyed by most men demonstrates that gender generally compromises the solidarity which has come to symbolize class-based resistance to coal companies for generations of coal miners and their families.

MINERS' GENDER AND CLASS CONSCIOUSNESS

During the economic boom of the 1970s, the class consciousness of miners was steeped in a masculine-identified "toughness" defined as a miner's willingness to defy management's authority by "standing up to the boss" in defense of one's rights (Yarrow 1991). A "good union man" was defined by his will to confront the company and, indirectly, the government. At the same time, this "toughness" has actually undermined class solidarity because it excludes women and fosters competition between male and female miners. In spite of these negative consequences, the conflict between the prerogatives of male privilege and the goal of class solidarity has made miners' consciousness dynamic and subject to change (Yarrow 1991). Since the 1970s, the class consciousness of Appalachian coal miners has been altered by the economic exigencies of the coal industry and the entry of women into the mines. Generally, miners have stopped consistently contesting managerial control and begun grudgingly making concessions. During hard times, they have generally "gone along to get along." For example, Yarrow (1991) found that during the 1990s miners showed a greater tolerance of "scab" labor than before. Ellen, who "grew up union," remarked, "They talk about scab miners and the nonunion people. I say them people got to eat and live just like I do. They got families to take care of, too." Moreover, the changes in miners' class consciousness show the ways in which the gender and class elements of miners' consciousness have coalesced. Their gendered belief system has been transcended to some extent, while class solidarity has persisted. Yarrow reasons that an "emphasis within the set of themes may change without completely disrupting accepted truths" (1991, 306). For example, at the national level, the CEP women miners' fight for parental and family leave in the 1980s and their participation in strike activity transformed a gender issue into a class issue (Seitz 1995). Nationally, the women's assertion of "sisterhood" was simultaneously a demonstration of "brotherhood" within the union. Locally, at the mine, although there were still traces of masculine-identified defiance, there were also signs of this emerging coalescence in both women's and men's class consciousness.

The resilience of the macho mentality was demonstrated by many miners in the Appalachian coalfields and beyond during the 1989–90 Pittston strike. This strike showed that union militancy and confrontational strategies had not disappeared, and that neither had the view of coal companies as the enemy (Couto 1993). Studies have shown that the women unionists who were directly involved in the Pittston protests were eventually assigned to less active roles and thus that their participation in the strike was provisional (Moore 1996a; Yarrow 1991).[6] At the mine I studied, however, the women and men miners were not only encouraged to join the Pittston picket lines but, according to some, were coerced into participating. A few women said they did not go because of the potential violence. Fearing the worst, Sadie chose not to join the Pittston pickets, saying that if she got hurt and couldn't work, she could no longer provide for her family. As with many men miners, her breadwinner role discouraged militant action (Yarrow 1991). In lieu of her direct participation, she contributed money to the strike fund, but not without negative consequences. "I got a lot of flack from a lot of the women and the men. It was said, 'What are you going to do when you need the union and they're not going to help you?' [The union president said,] 'If you don't go to Pittston with us tomorrow,' [then] when you needed him, he wasn't going to be there." Carla felt it was her duty to join the strike, but she resented being pressured to do so. "If you didn't go, you was the scum of the earth," she said.

Among those miners who went to the Pittston picket lines willingly, there were some distinct gender differences in how they perceived their experiences. To some of the men, the Pittston strike was about danger and confrontation. One recalled, "We went up to a machine gun nest. Pittston had machine guns posted up the hollow at the mine and we went up there carrying ball bats. I'd been with guys that shot," he said, pausing abruptly as if he feared disclosing too much. He continued, "Well, it's a wonder we all didn't go to jail over it. Just been lucky, I guess." His wife, who was present for this part of the interview, added, "They'd all been drinking." But, for the women, the Pittston strike was mundane and to some degree pointless. The women said they spent their time riding around in their cars. "We mostly just rode around in

6. Women became the first to protest nonviolently during the Pittston strike, when thirty-seven former women miners and miners' wives and widows occupied the front offices of the Pittston Coal Company's headquarters in Lebanon, Virginia (Green 1996). Believing that they would be arrested and jailed, they vowed not to give their names, and each woman planned to declare herself a "daughter of Mother Jones." After making headlines throughout the state, they left the building the following morning, victorious. But as Moore (1996a, 511) points out, men unionists felt upstaged by the women, who were subsequently and "deliberately excluded from strategy discussions" that led to the men's takeover of the Moss #3 plant.

the car all day. That's all," said Carla. "Just following a big line of cars," Lynn recalled matter-of-factly. "I went down there and rode around. I got nails in my tires one time. Pittston even throwed those jack rock things where you could get on them. A lot of us got flat tires like that." And Ellen demonstrated her cynicism about the strike when she said:

> I went there and we rode around. We seen people get arrested. They took some contract that everybody didn't want. Went off days of work for nothing and it was just a big show to me. Them people got the same thing that they was picketing about. I mean, everybody done all that for a big front. They adopted a contract they didn't want. We voted a contract down and they actually give it to us. So there's not no union no more like John L. Lewis and all that garbage, which, you know, maybe is better. I don't believe there should be no people killed.

For the men, the strike was about defending their rights as miners and their dignity as men, and their attitudes echoed the combative class consciousness of a bygone era. Fusing masculine consciousness with class consciousness, they expressed optimism about the strike's accomplishments. But the women were relatively unimpressed by the men's combative approach. They tended to see the Pittston strike as accomplishing very little toward changing the course of labor-management relations. They remained more focused on bread-and-butter issues, namely, contractual changes in the terms of miners' employment. In the context of miners' contemporary class consciousness, the men's "get-tough" stance has actually made the union weaker for several reasons (Yarrow 1991). First, law-breaking tactics, such as those described by the baseball bat–toting miner, only make men unionists more vulnerable to capitalist reprisals. Second, the union president's undemocratic and intimidating approach to soliciting miners' participation in the strike only led to their resentment and fostered feelings that their leaders had betrayed them. Miners' support for this style of leadership has proved to be counterproductive because it diminishes leaders' accountability to the rank and file and weakens class unity. Third, to the extent that men's narrow definition of "toughness" is associated with "manliness," it devalues other forms of strength and alienates the women who should be their natural allies.

But I also found that as the women negotiated their own identities as unionists and formed their own strategies for class-based resistance, they incorporated some elements of the men's more masculine-identified approach. For example, several women voiced a "toughness" sentiment of their own, making such comments about the company as: "You have to stand up for your own rights. If the women don't stand up, they will do you that way." Moreover, like the men, some women's tough defiance of the company underground had earned them the "troublemaker" label among company personnel. A "troublemaker," by company standards, was a "good union man" according to most miners. In this context, the "troublemaker" role was admired by most of the women, flirted with by others, and whole-heartedly embraced by a few who wore it like a badge of courage.

The coalescence of gendered elements in miners' class consciousness can also be seen in "the recent change of emphasis in the definition of manliness from 'standing up to the boss' to 'hard work,' which has accompanied the demobilization of the rank-and-file movement and the crisis in coal employment" (Yarrow 1991, 306). This adjustment for the sake of economic survival has allowed women to legitimize themselves as workers and as unionists when they cast themselves in the traditionally masculine breadwinner role. For men miners working alongside women, the breadwinner role appears contradictory, because at home men have control over women. At work, by contrast, they share power with women. While the women miners I interviewed readily acknowledged the men's need to support their families, they also successfully used the breadwinner role to assert their right to work underground. Over time, many of the men came to acknowledge the women's "breadwinner" status and right to work with them underground. "I think that the ones I have worked with," said one veteran male miner, "they have to work because both of them are divorcees. They have no other income." In this regard, women and men miners were united when management threatened their job security and benefits, and thus their means of providing for themselves and their families.

Clearly, issues such as women in mining and loyalty to family before union in hard times threaten the brotherhood that has been the traditional foundation of class solidarity. But the resolution of these conflicts represents signs of the union's emerging strength. The revival of miners' militancy, evidenced by the Pittston strike, and the coalescence of women's and men's class consciousness are encouraging developments. Despite the issues that threaten to divide miners, the UMWA has continued to be the major institutional form of miners' collective resistance. Historically, it has frequently awakened miners' class consciousness and refreshed their collective memory, even as miners pursue their individual interests (Couto 1993; Laslett 1996). As long as it serves miners' strategic and practical needs at the point of their intersection, the union will survive.

Using Cockburn's (1983) conceptualization of the relationship between gender and class, Yarrow has pointed out that the lived experiences of coal miners represents a fusing of both gender and class experiences. "Attempts to understand class consciousness in isolation of gender," he writes, "ignore an important element of consciousness that powerfully shapes action, for work and other aspects of life under patriarchal capitalism are structured by gender as well as by class" (1991, 285). Among the miners I interviewed, union solidarity persisted despite gender-based divisions. Women and men miners at the mine I studied demonstrated "a continuing sense of class conflict largely because they have continued evidence that coal operators often do not act in their best interests. Therefore, they see a continuing need for the union in spite of its present weakness and past mistakes" (Yarrow 1991, 306–7).

But miners have also come to recognize their own relative powerlessness in the workplace because of elements both within and beyond their control. Moreover, the women realize that the men's beliefs about male superiority and male bonding support a false consciousness that can be used by management to weaken solidarity among miners. As Yarrow (1991) has argued, these contradictions need to be exposed, because as long as miners remain divided over gender, management will continue to exploit them. Despite the contradictions in miners' class consciousness and their generalized despair, they still believe they have some measure of control through a common recognition of their own unity in times of trouble. For better or worse, they realize that the union has been and will be their sole source of collective support in their continued conflicts with capital. As a result of patriarchal practices, the women miners in particular have also come to realize that they can rely on each other. As several of them agreed, without the support of the union, that is all that's left. Their own sense of sisterhood, as they became socialized to the male-dominated world of mining, is the subject of the following chapter.

4

OVER THE LONG HAUL: ACCOMMODATION AND RESISTANCE TO THE CULTURE OF COAL MINING

During the course of their careers, the women were constantly confronted with the conflicting expectations of being women and being coal miners while working and interacting with their bosses, male co-workers, and each other. They knew they were seen as intruders disrupting the masculine-identified culture of mining, and to varying degrees they resisted men's hostility and sexual harassment. Their greatest challenge was knowing when to resist and when to accommodate to these situations. Their survival as subordinate group members depended on their knowledge and occasional use of the language and manners of those who dominated them (Collins 1990). In defining others, they redefined themselves as women mining coal. The accommodations gradually became mutual. Over time, as the women and men became accustomed to one another, they engaged in gendered interactions that resocialized them both. Although such behaviors as sexual jokes and profanity indicated that men still related to women in sexual terms, the women themselves said they often used such devices to establish bonds between themselves and their male co-workers.

THE WORK CREW AS FAMILY

Previous research on miners' social relations has shown that a miner's primary identity is with his crew (Vaught and Smith 1980). The women

I studied often spoke of wanting to stay with their crews because of the familiarity, loyalty, and sense of belonging that had developed between them. In Ellen's words:

> I got used to working with a certain group of people. Now, it's terrible to be that way and we're like family, it's just like we go there with our own section and we're not around nobody else, we're just around each other. I don't like the idea of being changed off and working with other people now. And that's sick, I think the company's got us like that. [*Laughs.*] Then if you think they're going to take you off and make you work with other people, you go, "Well, I'm not going to go on another section working, that's my section," and I've caught myself and other people saying that.

The interdependent nature of their work and the long hours miners spent working with one another fostered certain familial styles of interaction between the women and men at the mine. "These guys here and women here, they are your family," Carolyn said. "You see them ten hours a day if not more. You go home and sleep. Then you're up and come back again six days a week." Indeed, all the women likened their crew membership to being in a family, with all the support and cooperation that implied. "It's like you're just one big family." Lynn declared. "Everybody's working to help each other. If you don't, it makes your job hard. When you get on a section where people aren't like that, it makes your job hard."

As we have seen, miners admired each other for their ability to work hard, to observe safe work practices, and to demonstrate "team spirit." The women told numerous stories about male co-workers who had "looked out for them," which made them feel valued. For example, one miner's quick thinking spared Sally serious injury or possibly her life.

> There was this guy and he grabbed me by the back of my shirt and pulled me. I thought he was just goofing off. I was shoveling and I was getting ready to put a timber there. He grabbed me by the back of my shirt and he pulled me. A big old slab [had] come down and fell in between the two pins. It rolled down my leg and just bruised it because he had yanked me hard

> enough and far enough. [Otherwise] it would have fell right on my head and slid down my back.

On another occasion, Lynn, who had suffered a serious back injury underground, said she wouldn't hesitate to ask for the men's help because, as she explained, "I work with an awful good crew. I try not to [ask for help], but I've hurt my back three times and there are some things I cannot lift. If I think I need help, I'll ask them." Both women and men also voiced their disdain for conflicts among workers underground, especially when these rifts hampered solid working relations. Lynn said she "knew about two women having an argument in the mines. The men then said, 'You all should try to work that out and not have friction,' because they work together everyday." Whenever that occurred between any of the miners on her crew, she said, "the men [would] try to talk to them and [get them to] apologize or whatever has to be done, because you're going to have to work together and it can just get worse." Some men counseled the women about how to handle difficult men and bosses. "They'll say, 'Go to him and tell him this or that,' or how I felt about it, and get it settled," said Lynn. "Don't go ahead and work under that. It puts you in a bind when you think somebody's got something against you." Even some miners' wives recognized and respected the bonds between their husbands and the women with whom they worked. After one of her crew members had back surgery, said Lynn, "I called his house to see how he was doing. His wife talked to me and she said, 'Call anytime you want to.' I said, 'Well, I was a little leery about calling because I didn't know how you felt, but the section was wanting to know how [the injured miner] was getting along.'"

These underground "families" also functioned as a means of social and emotional support that enabled crew members to cope with the stress caused at work and at home. "We all have a bad day every now and then, and we need each other to help get us through them," said Kelly. "I think we all [feel] basically the same because we're always concerned about how our children are doing, or nieces and nephews. Every day you ask about how's [co-worker's daughter] or how's so-and-so. At Christmas, you can tell, because it's little gifts passed back and forth or cookies brought. We all know each other's birthdays." The same kind of support applied to the women's and men's problems with their own families aboveground. Lynn revealed:

> I'm a talker. Any crew I've ever worked with, I get to know the guys real well. I guess all the guys know all my problems. They've even told me, "If I was you, I'd do this or do that." Like I said, we're just like a big family in the mines because you're together more with the people there than you are with your people at home. There are just so many problems, sometimes I feel like my head's coming off. Then you've got to go to work and deal with everybody up there. There are a lot of family problems up there because everybody's always talking about things like that. If something happened bad at home, they'll talk to you about it just like you're family. I've always been close and I get to know a lot of people.

Men also confided in the women about problems at home. According to Michelle, "A number of wives would be surprised to know what I know about their relationships. It's the truth. They would be absolutely shocked, really." Indeed, the women, at least, had become so accustomed to this way of relating to crew members that they were uncomfortable in its absence. Michelle said that one time she "was working a section with one group that did not talk to each other for the whole shift. They didn't talk to each other or nothing. I said, 'What is the matter with you all?'" But relations with the men had limits. For most of the women, their friendships with the men were far more important to them than sexual relations with any one of them. As Renee put it, "We've worked together years and years and years. I'd rather be your friend than your lover. I couldn't do that. You become close to people. They bring their problems to work, they share it. If you have problems, you share it. You become involved with people, and it is easy to do that. You need to have respect."

At the same time, within the colonized Appalachian region, the family has endured as a patriarchal social unit in which gender relations and women's subordinate status have been slow to change (Hartmann 1984; Scott 1995; Seitz 1995). Thus, the family is the social context in which women learn about domination, accommodation, and resistance, three common forms of agency within a capitalist patriarchal structure (for further discussion, see Seitz 1995). While the women at the mine were accepted as workers, they were still informally recognized by their gender. For example, underground coal miners are famous for awarding each other nicknames as part of the informal bonding process (Ross

1974). Two of the older women I interviewed said they were given nicknames indicative of female kin. "They called me Granny," said Irene. "My daughter who worked there stopped that. 'She's not old enough to be your granny, so don't call her that.' She scolded them a little." Irene's daughter, like many of the younger women at the mine, recognized the implied devaluation assigned to older women like her mother. Clearly, she found the presumption offensive. The men on Lynn's crew nicknamed her "Ma." "I'm the oldest one up there besides our miner man," she said. "All of them seems to look to me that I'm older than they are, and they're a lot younger than I am. So they just carry on all the time like that, but they're a nice bunch to work with. It don't bother me." Although both Irene and Lynn reluctantly accepted these familial role expectations and styles of relating to their crew members, other women miners more readily understood the gendered implications of being "family" and how they were constrained by them. In our conversation about the solidarity among male miners and the women's sense of inferiority via internalized oppression, Ellen explained that "it all goes back to childhood—that the brother is better than the sister. The boy, he's supposed to be so much better. He's the boss and the girl is just a girl. She ain't supposed to be nothing. That's how I feel they feel, the way they act."

Like most women working in male-dominated occupations (Enarson 1984; Kanter 1977; Walshok 1981), the women miners had learned over time to pick their battles. Generally speaking, the informal norms of the workplace continued to be the men's norms and styles of relating to one another. "It's a man's world," said Kelly, "it really is." And to varying degrees, the women had conformed to some of these norms of everyday behavior. Noted black feminist theorists Collins (1990) and Lorde (1984) have discussed the effects of oppression on black women. They assert that black women learn to be "watchers" of those who attempt to dominate them. This watching generates a dual consciousness whereby members of the subordinate group come to understand their world from both dominate and subordinate perspectives. In an attempt to conceal their own view of things, they often adopt their oppressors' language and manners, which affords them at least the illusion of protection from them. From behind their masks, they are better able individually to resist their oppressors' attacks on their personhood and dignity and to deny their oppressors' definitions of themselves as inferior. I found similar processes at work regarding the women's use of profanity and jocularity as they interacted with men underground. Nonetheless, limits were set on the women's behavior, either by men or by the women themselves, which reset gender boundaries that maintained women's gendered status. These same processes also affected the women's beliefs and behavior toward one another, a subject addressed below.

Profanity and other taboo behavior is endemic among workers who engage in dangerous, anxiety-provoking occupations such as mining. In addition, the mine itself is an "encapsulated enclave" that is conducive to such behavior (Vaught and Smith 1980), and coal miners are notorious for using foul language underground. Company rules forbid the "use of profane, obscene, abusive or threatening language or conduct towards subordinates, fellow employees, or officials of the company." Swearing directed at another individual was considered "conduct which violates the reasonable standards of an employer-employee relationship." Needless to say, the swearing rule was consistently violated by women and men alike, depending on the circumstances. Swearing in general was usually tolerated and sometimes followed by an apology. But swearing at another individual was not. The women at the mine varied considerably in their use of foul language and their willingness to tolerate other miners' profanity. Some women were very tolerant. As Michelle remarked:

> A whole lot of people say that they mind cussing, but I don't mind cussing. I look at a coal miner's job like a policeman's job, when you step into it and you don't know if you're going to come out of there alive that day. I listen to some of the men here talk. Their wives are bitching with them at home, the kids are screaming, and when they come to work most of them are working for a boss that they don't like. Then they have union people hassling them. I give them a little leeway. I don't care if you cuss, as long as you don't cuss at me. I don't care what you say, because you can't say anything that I haven't heard before. So I don't object too much.

Sadie had a very different opinion.

> I can block it out. I did tell them, "I don't expect you to change your vocabulary for me. But," I said, "don't direct your four-

> letter words right at me. I don't want that. Now, if you cuss me and use your filth," I said, "I will do something about it. If you want to talk the way you always talk, talk in front of your wife and kids like that, go ahead and talk that way." I said, "I can block it out," and I did.

Regardless of how they felt about men's use of foul language, none of the women was willing to accept the men's profanity when it was aimed at them. Sadie told me, "The boss cussed me on a mine phone and I went to the safety man and talked to him [about it]. They made him apologize. Maybe if he done it to my face and not on the mine phone where everybody could've heard, I would have let it go. But I didn't that time."

The women also said that some men would apologize if they thought a woman had overheard them. Their apologies implied that there was a difference between men's and women's language. Specifically, if profanity was not fit language for a woman to hear, then it certainly was not language she should use herself. Kelly explained, "I know some men watch their mouth. Because they say the big f-word sometimes. The language underground is terrible. And if they see one of us standing there, they'll say 'Excuse me, I'm sorry.' But that's an everyday word for those guys. So, like I said, the language underground, and even my language underground, is totally opposite to what it is here. So we accept each other for that." Jessie said that although her language has gotten worse since she started mining, she doesn't "cuss and carry on" at work, and men's apologies for swearing in her presence make her feel good. Irene, now retired, declared, "I could talk just as much trash as they could, but I didn't know many filthy words." Sally, who had retired after a serious back injury, elaborated. "Those coal miners got the foulest mouths of anybody you ever heard in your life. Well, I didn't cuss when I started to work in the mines, but when I left I cussed. . . . I really didn't know I was doing it. It just fell in. It wasn't like you took your part or anything. It's just like you automatically done it and didn't realize it."

For other women, though, using profanity was out of the question. Although some admitted using "men's language" underground, they were careful to restrict or conceal their use of profanity. Some women swore only when the situation between them and a male co-worker warranted it. "When you're really angry at a man, where you come to the point where you got to cuss him out," admitted Lynn. "I have a real high-strung temper and if I get mad, the first thing I'll do is cuss, and sometimes it might not be so nice. [*Chuckles.*] I mean, I don't make no habit of just going around cussing all the time. I have to get pushed pretty far before I start really talking bad." Other women who used profanity were quick to hide it. Ellen said, "I used to not cuss too bad, but I'll cuss now. [*Chuckles.*] I'll say it under my breath. I don't think they've ever heard it. They'd die if they heard me say what I say to myself because they would think, well, that ain't Ellen." Finally, on those occasions when women did swear around certain men, the women extended apologies of their own. As Irene said:

> I cussed a boy one time. We was pulling a water line through a stopping door. It was muddy and this boy [was] behind me, pulling. He took his hand and patted me on the shoulder with all that mud on him. I said, "You dirty son of a bitch." Oh, he made me mad. I couldn't talk to him the rest of that night. He laughed. I didn't think it bothered [him], but I went to him the next day when I got to work and I apologized to him. I said, "I shouldn't have said that. I know you're mother's not a bitch. I have no hard feelings against you."

Carolyn confessed, "One time I cussed in front of my boss, who's a Christian. I mean, I cussed bad. I mean, I just lost it. And I said, 'I'm sorry [foreman's name]. I'm really sorry.' I shouldn't have cussed because he's a Christian and I respect him. He knows I'm not a Christian and he knows I cuss some. But if he's around, well, it's out of respect for other people." She continued, "I cuss some when I get mad, but I always try to watch what I say because of, I'll lose that edge." I asked what that "edge" was. She replied, "Respect." The norms governing the use of profanity had become reciprocal between women and men at the mine and represented a mutual accommodation that preserved sexual differentiation underground. At the same time, the women's acquiescence was relative and represented a form of resistance that afforded them a measure of dignity while they worked.

Jocularity includes pranks, teasing, and telling jokes. Although company rules forbid "horseplay," miners still pulled pranks, told off-color jokes, and delivered mild insults to each other. Like profanity, jocularity reduced the tension of working in the dangerous mine. Jocular behavior also promoted social integration and cohesion among crew members. A few of the women talked about being "initiated" into their crews. Sally explained:

> There's what they call initiation, where they whip you with a wedge. But I didn't get whipped with a wedge. I got whipped with a board with nails in it. They took the nails in this board and they bent the nails over. Burnt me up with it, the mechanic did. I never known them to do that to a male miner, just picking on the women. But the guys I work with? It was just in fun, they wasn't mean to me.

She also told me, laughing intermittently, about another incident.

> They took my boots off and threw me in the break. Took my cap light and my boots and I was just sitting there in the break, thinking, well, I guess somebody will come along. The one that done me like that come back and got me. But that saved me from unloading that motor. So he had unloaded all that stuff and came back and got me. So really, I just sit in the break with no lights and no shoes. And he went and unloaded the flat and then when he got done he came back and got me.

But, she added, "the only reason I wasn't left there long was they seen the boss coming. Man, they cut me loose and give me my cap light back."

When Lynn started working underground, a woman on her crew "had the tendency of talking with the men and talking with the boss. It would keep her out of work." She said her crew, who "was crazy [because] they carried on all the time," told her, "'We're going to make a coal miner out of you.' They would dump the coal on the rib for me to shovel it. You ain't going to be like that woman over there. I didn't have much of a choice. I had to learn to work hard." Although the women could have reported these incidents to company personnel, neither one did. Their belief was that the temporary loss of individual dignity afforded them more lasting acceptance among their all-male crew. Paradoxically, however, by bending to the will of the crew, the women also achieved a sense of their own individual identity and developed their own definitions of themselves as coal miners.

Most of the women believed they had good relations with the men.

They said that they liked to joke with them and enjoyed the camaraderie. For example, Michelle, who described herself as "Whoopi Goldberg II" because of her propensity for making wisecracks and telling jokes, told the men on her crew, "You would never meet a woman more open. If you really talk dirty, I can keep up with you. If you want to talk about preaching, I can talk about that, too. You can talk about anything you want to talk about around me." Lynn spoke of the spirit with which the men on her crew related to her and to each other.

> The bunch I work with is always telling me they're going to whip me and all I ever say is, no. I'm bigger than they are! [*Laughs.*] They just play around. It makes their night go by and you've got to have a little bit of a sense of humor. It's a little bit rough. One guy that runs the miner, he'll say, "Don't you all bother her, she'll whip you now!" They just carry on with you all the time. They're just joking. I take it as joke just like they do. They're a good bunch to work with.

But not all the women said they were initially comfortable with this style of relating to the men and that it took some getting used to. "I was never like I am now. So that took years to get me there," said Kelly. "Now, when I first came here I was totally no joke, nothing, and it was hard. I took jokes hard and I had some hard feelings, but I loosened up after three or four years. Now my workplace—I enjoy it." After twelve years underground, she characterizes herself as a "prankster and a joker," adding, "That's how we all get through our day." Still, some women resisted men's offensive jokes and remarks and used silence to feign acquiescence. "They say something to me once in a while, but, you know, I act like I don't hear it, or I act like I'm dumb and I didn't understand," said Ellen. A few of the women never made these social adjustments at all, and they consistently complained about the men's behavior. The other women regarded them as "crying foul" or "playing the victim." "Some of them take everything too personal," said Lynn. "They think the men are really trying to be rough on them, when, after they get to know you, most of the time, they'll joke around with you just like they joke between theirselves." After all, she said, "You've got to get along with them. You can't buck up against them because there's too many of them there, and they was there before we ever got there. If you buck up against them, they're just going to cause you more problems."

Those women who could "act like a man" and still "be a woman" demonstrated considerable inner strength. This was their accommodation as well as their resistance. It was inconceivable to them that a woman miner seeking legitimacy would behave any other way. Once women adjusted to the masculine-identified norms of mining, they also enjoyed certain privileges of reciprocity and rapport with their male coworkers. But regardless of gender, when it came to jocularity, said Kelly, "You've got to be able to take it in if you dish it out. You can't be totally dishing all the time." Lynn's lengthy story below illustrates the point, but it also demonstrates how time after time women learned to "get along" and even be appreciated by the men with whom they worked.

> [A younger male crew member] started calling me "Henrietta." Then they started calling me "horny Henrietta." The mechanic added the other part. He was a smart aleck and he got the filthiest mouth of any man I've ever seen in my life. But now it shocked me. I come home and I told my husband. I said I like that section, all but that mechanic. Then some of the men gave him a nickname not too long ago. They started calling him "Paunchy." When they told me, I rode that in the ground. Every day, I hollered, "Hey Paunchy." He got mad and went to the boss and he told him that the men was calling him that name. It burnt me up, and the men on his section. See, he didn't tell on me, he told on them. Now those boys aren't allowed to say anything to him. They told me they couldn't call him that no more. And I hollered at him, I said, "Hey, Paunchy." He turned around and I said, "Go head and say what you want to me, old buddy, because I want you to tell me I can't call you that." I said, "You're the one that started that name on me up there. And I'll call you Paunchy until the day you die." I said, "I want you to take me to the boss." I said, "I bet he'd like to know what you call me." He ain't said one word to me since. He don't even answer me when I call him that. I said, "I can't believe you're that big of a cry baby." I said, "You better go home to Mommy and get you a bottle." He could dish it out, but he couldn't take it when somebody got one on him. Well, all those men started laughing when I told him that, because they couldn't do that. Boy, it tickled them to death when I told him he wasn't nothing but a baby.

At the same time, however, the social accommodations that marked the men's acceptance of the women also preserved sex differences underground because they were established on the men's terms with the women's cooperation. Kelly's account of how she and Alice behaved toward their male co-workers illustrates the point.

> See, I get along with the guys really, really good. We harass each other. The men don't really harass us, we harass the men. I'm making contact with fifteen to twenty guys a day. And, like, on a supply move, forty-five guys, you know, and two women? They love it having us up there because it's a different atmosphere. We joke more and, well, we're a little bit flirty, but teasing, and those guys love it. If anybody's harassed around me and Alice, it's the guys, as a joke.

Although seemingly innocuous, flirtatious behavior allowed men to reassert their masculinity in the presence of women who otherwise threatened their notions of themselves as miners and as men. Once the initial discomfort of having women in the mines had passed, the men became familiar with the women in ways that accentuated their femaleness. For example, during our interview in the women's bathhouse, Kelly and Alice told me that, like their male relatives, the men they worked with knew when they started their menstrual periods.

KELLY: And believe this or not, [and] I'm only speaking strictly for myself, they know the minute . . .
ALICE: The *minute!*
KELLY: The minute you start.
ALICE: Right. You start. I don't know why. They come around and say, "What is it? The visitor from hell? The visitor from hell is up here!" They call it the visitor from hell when you start.
KELLY: They try to help us see the humor in it.
ALICE: But it's awful. [Pauses.] I love my crew, I do.
KELLY: They help us a lot.

Regardless of the extent to which these coal-mining women became socially integrated underground at the mine, the men still maintained their dominant status via the "sexualization" of work relations. Moreover, the women also became aware that, in their absence, male miners continued to do "crazy stuff" that indicated that they were "in their own world," unless, of course, they were caught off guard. Renee, who worked at the face on the plow, recalled such an occasion. She walked up to see "this one guy [who] was mounting the other guy." She said, grinning, "I looked at [him] and he was just wounded. He was more embarrassed than I was, and now every time I seen him. [*Laughs.*] But that's the way they get by. See, there are not enough women there, and they forget that they're there. They've worked for years without a woman on the section. Then one comes up there, and they see somebody walking they assume it's one of the guys." In a similar situation, Jolene decided not to make her presence known.

> I get migraines real bad, and one day I got one. It just got so bad I finally told the boss, "I'm just going to have to go out," and he says, "I don't have anybody to put in your place. Just go down to the end of the track, get in the bus, and try to go to sleep for thirty minutes and maybe it will go away." So I go down there and I crawl way over into the bus and I'm lying there. Some men come in and the first thing one of them does is use the bathroom and it's falling right down into a hole I can look right down into. I'm sitting there holding my breath. I said, "Oh God, please don't let me cough or anything." Then they took out a tape measure and they was measuring stuff and they never did know I was there.

As with their use of profanity, sometimes the men restricted themselves when playing pranks or telling off-color jokes in the presence of women co-workers. "If they tell a joke or something," said Lynn, "I'll laugh about it and then just go on. They don't ride it no farther, they just let it go. They'll say, 'Have you heard this joke? You care if I tell this joke?' And I say, 'Go ahead and tell it.' I don't care because I just listen and go on. You'll get along better if you can take a joke." But, more typically, as women have entered male-dominated work settings, they have often found themselves obligated to set limits on men's activities in order to avoid being degraded by them (Gutek 1985). At times some of the women miners also found themselves in the position of having to draw the line on men's behavior they deemed unacceptable. For example, sometimes Carolyn attempted to curb the men's "sex talk." "When they start saying a lot of different things, I say, 'Hey, you all shouldn't talk like that, watch your mouths.' They would start making sexual remarks about their girlfriends and women and I'd say, 'Hey, you shouldn't talk like that—what's the matter with you guys? You ought to be ashamed of yourself,' and stuff like that, just to get them to watch what they say."

Jessie said that although she generally "laughs stuff off," she was careful not to "get rowdy" with the men on her crew because invariably things would escalate. First, words might be exchanged between the men and a woman, she said, and "next thing you know, one of them is hitting you on the hind end with something. I know some women that just hits them back," she said, but that was not her style. In fact, she commented that occasionally if they got carried away, she would remind them, "to make them stop." She concluded that if a woman wanted to "get nasty with them," most of the men were willing to oblige them. Other women's stories corroborated her statements. Jolene told me, "When I first came here, they sent me running buggy for a boss. They said, 'We whip everybody that comes on this section.' I said, '[You do that and] you're going to be in the mine office'—and I would have. I'm not for that . . . and they're not going to do it to me. It goes on between the men. They take their clothes off and everything if they get married or something." And Sally said:

> Most of the time, when you're a mason, they reroute the air and you go back in no man's land, as I call it, and build stoppings. There was a grease bucket back there and one of them says: "We're going to grease you," and one of them grabbed me. I said, "No you're not either. The first one that touches me with that grease, I'm going to take this metal on my boot and see how hard I can kick them in the head. Now, if you can hold me down and grease me, go ahead. But when you do, it's the first thing I'm going to tell when I get outside." So me and the guys, after that, we always got along.

As Carolyn commented about the men in general, "You're not going to change people." But by calling the men's attention to their behavior, she concluded, "All you can do is have them have respect for you." By setting limits on their own acquiescence, these women opted for a more active form of resistance to the men's gendered behavior.

SISTERHOOD

Over the years, the women working at the mine discovered that one potential source of their strength was their own solidarity. However, it is important to remember that within the matrix of domination, those who are oppressed can also be oppressors; no one group is completely privileged or penalized (Collins 1990). One reason why socialist feminists do not view women as a class is that some women are oppressed by other women. Moreover, men's subjugation of women is not always uniform, and male ideology is not as cohesive as one might expect (Seitz 1995). At the mine, some men accepted and assisted the women with whom they worked, while some women, who apparently embraced elements of patriarchal ideology, assisted other men in "otherizing" women. Because the women used the men's work standards to evaluate and compare one another, their evaluations were based on each other's willingness to work hard and act responsibly toward other miners. But their estimations of each other were tinged by racism and by perceived differences in women's physical attractiveness. Specifically, black women and more physically attractive white women were believed to receive preferential treatment on the job. Although these perceptions had softened over time, they still hindered women's collective resistance. When collective resistance is not possible, individual resistance tends to occur (Cable 1993).[1] So when the barriers to collective resistance were insurmountable, the women at the mine often resorted to their own individual strategies for resisting male dominance. The following

1. The term "collective resistance" was originally developed by Stephen Lukes (1974) in his discussion about quiescence in the power relationship between dominant and subordinate groups. Gaventa (1980) applied the concept to coal miners in a central Appalachian valley in explaining their relative lack of collective resistance in the face of three types of power routinely used by elites to exploit and impoverish them economically as well as culturally. But as Cable (1993) points out, a lack of collective resistance does not mean the absence of any kind of resistance. Where collective resistance is not possible, individual resistance can still occur. Cable explains that individual resistance not only tends to reflect the traditional rural value of self-sufficiency, it also reflects the historical effects of the coal industry on the region's residents. The coal-mining families and communities of the central Appalachian region were divided for decades because of the confining nature of the coal camps and the companies' attempts to promote competition between them. Sachs points out that rural women are more inclined to rely on individual everyday forms of resistance even while they ostensibly remain "in their place." She writes that "patriarchal forms and traditional values prevail and organized feminist struggles remain relatively rare in rural areas. Rural women shy away from widely embracing feminism, but this by no means suggests that these women do not resist domination" (Sachs 1996, 27).

section focuses on the conflicts and bonds between the women at the mine, what they thought of one another as workers, and how they related to one another on the job.

Racial Antipathies and Harmonies

Some of the white women I interviewed in 1990 said that black miners generally received preferential treatment because they were not required to work as hard as white miners. Sally mentioned that a black woman she had once worked with "got out of work a lot because she was black [and] she would threaten to file grievances" over the work she was given, although it was no harder than other women's work assignments. As a result, her boss was reluctant to assign any black woman the more arduous jobs. In turn, Sally told the boss that she was "going to file a grievance because all the blacks was getting off." At that point he started helping her with her work, even though he was contractually forbidden to do so. Sally concluded that because the company had been made to hire women and blacks, managers were doing their best to accommodate both. Citing the political power black miners' had on account of their race, Kelly said with a note of disdain, "I don't think they get done wrong because of the NAACP. They just call them down and they're here. That's it." She concluded that white women were the least supported group at the mine.

But what some white women miners considered to be management's preferential treatment of blacks was seen by blacks as racial sensitivity. Michelle, a black women miner, said, "I think [the company]'s problem is that they don't want people to get the perception that they're mistreating black people or black women over here, and that's pretty much a push for the white people not to . . ." she trailed off. "Mess with you?" I suggested. "Yeah," she replied, "they don't want to hassle about it." She said later, "I do what the boss tells me to. I tell the guys, 'If you have a problem with what the boss tells me to do or what the boss tells you to do, then you go to him. Don't you come in my face.' I said, 'Because he is all our boss. If you have something to say to him about different treatment, that he's making it easier on me than he is on you all, then you all go to him.'" Moreover, I found that the stereotype that blacks are lazy was refracted through the lens of gender. For example, Lynn declared, "Most of the colored women, they're not that good of workers. The colored men are the same way." But she did allow that a boss "would rather have a [black] woman than one of those colored mens." During their tenure at the mine, Michelle and Renee, the two black women miners I interviewed, believed that they had very solid work reputations. No one had ever said anything directly to them to the contrary, but they felt that if anyone did have this perception, it was because of their sex rather than their race. Nonetheless, Michelle did her best to negotiate her own identity among the men she worked with relative to her white sisters underground.

> I think guys like us because we're down to earth. The white men actually treat the black women better because the white women, some of them are stuck up and they won't talk to the fellows. I can get along with anybody. I can work with anybody. I think they like us the best. We come out there in the morning and everybody's hollering at us. "How you doing? What's up? Hey, girl!" You don't hear them saying that to many of the white women. They don't talk to the guys, and we do. I think the white guys would probably say we are friendlier and we talk to them more.

According to Collins (1990), relative to black women, white women represent the "cult of true womanhood." Underground, where a masculine-identified code of conduct prevails, Michelle reconstructed white women as the true "others" according to this image, which allowed her to claim greater legitimacy in redefining herself as coal miner.

But over time, instead of entrenched racial strife at the mine, I found more evidence of racial harmony and of contempt among most white women for generalized racism. A few of them were surprised when I asked about racial tension at the mine. If there was any, they avowed, they certainly weren't aware of it. Since the late 1970s, there had always been a few pairs of white and black women working together quite amicably. One afternoon I had a casual conversation with Jessie and Liz, a black woman Jessie had worked with for several years. Both of them believed that there was more cohesion than conflict among the women at the mine, regardless of race. Jessie, looking at Liz, remarked, "We stick together, don't we?" Likewise, Sally recalled working with Liz, and said, "Like me and Liz? We hung together. If something was done, like if she done it, I was with her. If I done it, she was with me. We stuck together, you know. Like these two others that I know. They stick together. So yeah, after you work together a while, you buddy up. You're buddies. One will stand by the other."

Jolene said she "had worked with several black girls, and we got along just great. They were the only ones that said the n-word, [but] they were just joking around." Like the use of profanity, which was tacitly governed by certain rules, the n-word was something that only blacks could rightfully use. While it represented the internalization of controlling images, it also set a racially based boundary separating blacks from whites. During my interviews, a few white women joked sarcastically about dating a black man or having a black baby at home. These remarks were made in reference to themselves for their mock shock value. Knowing that they were already stigmatized for engaging in gender-inappropriate work, these comments represented the women's mock acceptance of and true immunity to both forms of "deviance." Thus, in yet another instance, acquiescence became a form of resistance. At the same time, the jocular nature of their remarks does not negate the possible reinforcement of negative stereotyping that otherizes African Americans.

When I returned to the mine six years later, rather than stories about the preferential treatment of black miners, I heard numerous stories about long past incidents of racism underground. Generally speaking, the women I interviewed recognized racism more readily at a group and personal level than at an institutional one. They also vowed to confront racism when it occurred. Kelly's story about a discussion she had with her white male co-workers demonstrated her recognition of racial bias and its oppressive effects.

> We talked about O. J. Simpson. That was a biggie, a biggie. In this area there's a lot of prejudiced people. And it's okay that O. J. killed Nicole because Nicole was with a black man. That burns me up. Just one or two people would make that statement. I hate to say this word, but I've got to say it so you'll understand. He killed another nigger—what difference does it make? I'm like, what? Well, she married a black man; that made her, uh, [*pauses*]. Oh, please don't tell me you've got that mentality!

During the same interview, Kelly asserted her opposition to racism at the mine even more forcefully. "One day a boss made a statement and I really scolded him for it. He said, 'Well, I'm going to go up there and see what my little nigger boy's doing.' [Like,] 'You get a bunch of them and you got problems.' And you know that irritates me so bad. Like you say there is ten to one whites over blacks. Nine lazy whites you don't think nothing of, [but] one lazy black?" Kelly understood that blacks were just as much "categorized" as women were, because one black miner who was perceived as fitting the stereotype represented the entire racial group underground. Ellen, Kelly, and Sally maintained that when it came to working underground, race did not matter to them. "I can honestly and truly say," said Kelly, "I cannot see skin color with my black friends." But even so, while she meant to imply her acceptance of African Americans, her color-blind remarks still represented an unintentional validation of racially steeped images and beliefs about them (Collins 1990).

In their own running dialogue, Michelle and Renee demonstrated their understanding of the individual differences between them as miners. Renee told Michelle, "You go on the job like you're shopping to see what needs doing." Michelle retorted, "I know what I can and what I can't do. I'm not going to hurt myself." Renee then explained that Michelle had "a different outlook. I don't know why. [But] if I know I can do it, fine. But if I can't do it and they ask, 'Have you tried to do it?' No, I haven't tried it. I'm going to try and find I can't do that. I can't do it, but I tried." Laughing while directing her remarks to Michelle, Renee said, "It's a different thing from, 'That's too high. That's too heavy.'" They concluded that blacks and women needed to be seen as individual workers and not as stereotyped representatives of either group.

Overall, most of the women I interviewed demonstrated an awareness of how the issues of region, class, gender, and race intersected to compound the controlling images that subjugated them. Giving her assessment of race relations at the mine, Kelly said:

> You know how those guys are? They'll talk trash about you behind your back and eat you up to your face. So there's never a risk of violence. They would never say that in front of them. There's always little slurry jokes behind their back. And no man or woman up there has been given a job simply because they are black, because we have the seniority. That's how important it is to keep the union.

Similarly, Ellen remarked, "I can't see why, if a black man can do the job a white man can do, and they're both hired, let them both have the glory of doing it. I would stick up for any of them myself if I thought they was in the right. It doesn't matter what they are, just like the blacks, the women, the men." Later in the interview, she warned of the danger in allowing gender and racial oppression to divide workers who were all subject to the same class exploitation. "I believe it's more who you are anymore than if you're a woman, I swear I do. If you're black, I hate to say it, and you wouldn't believe the bosses' [racial remarks]. I mean, there's a difference there, but when you're down there working, [it doesn't matter] if they're men or women, [and] it ain't their color." Women miners recognized that classism and regionalism constituted a similar set of oppressive images and beliefs. As Seitz (1995, 215) has written, "In the sense that 'Appalachian' carries a particular meaning in American metropolitan culture, it is an ascribed ethnicity, a social construction that operates similarly to race in marginalizing and demeaning 'other' groups within the dominant ideology of the larger society." The compounding of class and region was captured in Sadie's remarks when she asked me rhetorically,

> Have you ever noticed when they interview someone on TV from West Virginia or somewhere, they get the people [who] look like they're from down in a holler? I don't think it sets a very good image for West Virginia, especially the coal miners. We're not all like that. It puts us in the class of the old, dumb coal miner, and we've been called that an awful lot. You can't really be dumb and be a coal miner!

The interlocking systems of oppression—gender, race, region, and class—get their strength from their victims' acceptance of the oppressors' controlling images of them, by which they become powerless to resist them either individually or collectively (Collins 1990). But the women's awareness of their oppressive effects allowed them to develop more accurate and positive self-definitions. To a degree, this awareness had become a source of empowerment by which they had endured, despite other gendered differences that at times threatened to divide them.

The Beast of Beauty

When women joined the ranks of coal miners, male miners acted to preserve the gendered status hierarchy in the workplace via the sexualization of work relations (see further discussion in Reskin and Roos 1987). They exaggerated gender differences in order to minimize the women's challenge to the presumably "natural" order between them as typically found in the patriarchal family. As the "natural inhabitants" of the coal mine, men were exempted from similar gendered pressures. Women thus were subjected to the men's assessment for "doing gender" (West and Zimmerman 1987). In other words, the women at the mine were held accountable for engaging in gender-appropriate behavior through both men's and other women's evaluations of them based on normative conceptions of femininity, including standards for sexual conduct and feminine beauty. And not only were women under pressure to be attractive to men, they were expected to prove themselves as coal miners if they expected to keep their jobs. These dual pressures made the women's advancement more difficult, particularly when other women deliberately stood in their way.

Women miners experienced conflicts with other women over jobs, work assignments, and other discretionary work conditions, such as getting off early or having an absence excused. All the women and men I interviewed mentioned these conflicts, which they believed were motivated by job-related jealousy. An older male miner said that, in the past at least, "I think you saw a lot of jealousy among the women. Like, say, Jolene's job. She has a kickback job where she doesn't really have to crawl around and get dirty. Well, you saw the other women resent her for that. Then it comes somebody else's turn and they all jump on her." Just as some women felt it was natural for men to pay more attention to women underground, some also thought it was typical that women were prone to resenting other women's success. Indeed, some women believed that women were somehow worse than men when it came to job-related jealousy. Sadie admitted with dismay:

> You've got petty jealousy among the women. You do a whole lot. It's over jobs. And you've got a better job than I've got, you've got an easier job than I've got. If you feel cheated, you're going to fight. If you get a whole bunch of women together working, maybe I'll have to say that women are a little more picky and petty than men. I think you would have a little bit of a problem there [with] a whole bunch of women. I know just in our bathhouse for a little while they're always picking, and it boils down to jealousy, mostly.

Even so, other women recognized that not all women resented another woman's success and that some men were equally threatened. In her own defense, Jolene told me:

> I can honestly say that the biggest problem I had here when I got my [fireboss certification] papers [was that] some of the women gave me a hard time, catty remarks. I work hard. There's people I work here with and lots of people here on other shifts I don't know, and they say, "She don't have to do anything, they don't make her do anything." You know those concrete blocks that have wires in them? I set them for two and a half years. They weigh fifty-eight pounds apiece.

When asked whether she would prefer to work with another woman or a man, Ellen replied:

> Well, some of them I would and some of them I wouldn't. Because, I'll tell you, when you work with another woman, like if one gets a good job and the other one don't, the other one gets jealous [*pauses*]. I'm going to tell you something. I love all of them, but ain't nobody can get along with them hardly because there's so much difference in them. Like, if I go in and I might have an easy day, which I ain't never, and they didn't, they probably pick at you or something. Sometimes I like to work with another woman, though, if I could pick who it was.

Other women understood that, like the men, women differed in their abilities to do certain kinds of work. Talking about two other women at the mine, Sally said:

> One of them run the scoop and the other was shoveling the belt. This one got mad at the one that was running the scoop and wouldn't even speak to her. She didn't know why she wouldn't even talk to her. Well, the one that was on the scoop could scoop better. She was a better scoop. So they took the woman that could do the better job and then sent the other one on down the belt to shovel. I didn't see nothing wrong with it because she wasn't by herself. She was down there with a bunch of other people up there shoveling. And she got mad at this other

> woman who stayed in the face and scooped. But she was doing her job.

It seemed clear to me that some women had at least partially internalized the men's stereotypes of themselves when they asserted women's propensity for jealousy of other women. Others showed a measure of resistance to those stereotypes by asserting, as Ellen did, that the women at the mine could not be categorized as a group.

Categorical thinking is simplistic, and those who resist it complicate the lives of other subordinate group members who have adopted it for their own protection. Structurally, when one woman stepped over the gender line, men—and even other women—held all women accountable for it. As several women who had been firebosses from time to time found out, other women were just as defiant of their authority as men were, if not more so. "Men don't think women should boss them, and the women sure ain't going to let you boss them," said Ellen. "See, I'm a fireboss, and on a Friday or Saturday we go downstairs and we had to search buckets for smoking materials. Them women [said,] 'You ain't touching my tobacco,' you know, like that. But I really could have made it rough on them if I wanted to, you know? They was so jealous." Jolene thought that the men were actually easier to supervise than the women because, she said, "When I got my papers, they gave me some women to work for me, they made some catty remarks, like, 'They must be hard up for foremen.'" These women concluded that there would never be a woman in a permanent position of authority at the mine, because such a person would be resented by men and women alike. Moreover, as Ellen said, "I wouldn't want to try it, because you'd be stepping on so many toes that if the union didn't get you, the company would."

Those women miners who had been the objects of other women's resentment surmised, as Jolene had, that "the ones that don't want to advance resent those that do." Similarly, Renee said, "Some of the women don't want another woman to have more than what they do. They don't want that. They want to keep you at their level. I guess they're used to that up there." Carla believed that "basically they wasn't getting the respect that you've got, so they're jealous and they find things to nitpick about constantly." But Ellen felt that the true basis for the women's jealousy was some women's fear of failure. "I made it," she said, "but they are afraid to do it. All people are afraid of things, and what you're afraid of, you make excuses about." In general, women who had developed a greater task orientation to their work were more aware of the men's stereotyping of women. They also tended to believe that women had the right to pursue whatever job they could adequately perform. "The women, now, have come a mighty long way," said Renee. "[But] they have a long way to go, too, as far as training wise. If you really want something, you're going to have to go for it. No ands, ifs, or buts about it. Don't stop. You can put it off, but don't stop."

In the contexts of the patriarchal family and a workplace that was similarly structured and controlled, some women at the mine had embraced certain elements of the men's negative controlling images of women. Rather than resist this form of oppression, they had internalized it. They were subject to men's stereotypes of them and had difficulty seeing other women or themselves as anything other than subordinates. They resented and defied women who had succeeded underground and were correspondingly less able to develop positive images of themselves as coal miners. However, for very different reasons, a few of those women who had advanced had also adopted elements of men's stereotypes about women miners. As tokens, women who succeed in male-dominated occupations such as coal mining have been regarded by men as "exceptional," while the rest have been "otherized." Some of these "exceptional" women in turn develop an "insider" attitude toward other women workers by adopting men's gendered stereotypes (Kanter 1977). "I think they want to feel as important as the men," said Ellen, "but I don't know how they're going to do it." Complaining about another woman she considered an insider, she said: "We had this one woman who would talk about the other woman [on her section] and get this other woman in trouble by saying she wasn't a hard worker." Readily admitting that she accepted what men said about other women, Lynn told me, "Most of the women, when they first started, didn't [work], and there are some that still don't." Her attitude not only exemplified an "insider" mentality, it also reflected the gendered relations of work underground that fostered divisions among the women who worked there.

Further complicating women's success underground were the men's externally defined standards of beauty. Both men and women said that a woman's physical attractiveness substantially affected her work status. According to Ellen, standards of beauty were imposed upon the women from the day they were hired. In 1980, when she applied for work, the company did not want to hire her. She called a management official at the mine to ask why. "He told me that if the foreman likes your looks," she said, "he'll hire you." Frustrated by his answer, she contacted the human resources director at the company's headquarters located in a neighboring state. "I asked him what was the requirement to get a job. He said, 'What are you talking about?' I said, 'Do you have to have a mattress on your back and be built like Marilyn Monroe to get a job in the coal mines down here?'" Later, I asked her what a woman would have to do to get promoted. She guffawed and said, "If it was twenty years ago, I was younger, and built real good and beautiful, I could probably get advanced up there!"

Race also entered into externally defined standards of beauty. Michelle said, "If you're a white woman there and you look halfway decent, you can about have your way. If you're an ugly white woman? They are treated worse than us." Renee reiterated, "If you're an ugly woman, you're in for it rough." The only way an unattractive woman could gain any measure of respect, she said, was to "stand up for herself" and actively resist her subjugation.

Conversely, miners of both sexes said that some women had used their physical attractiveness to further their ambitions. One of the men said that "the pretty women were treated better than the less attractive ones. If they was heavy set, they were sort of ridiculed." The more attractive women would get easier job assignments and tended to have a better chance for advancement, he said.[2] Several of the women agreed with him, saying that attractive women were given more training opportunities than other women. As a result, unless their work reputation was firm, attractive women were viewed with suspicion and resentment. As Lynn put it, "I think the majority of them are jealous of one or the other because as I said there's certain ways that some can get some things and some can't. I think it causes a lot of friction in the women."

While most of these tensions among the women were sporadic and short-lived, there was one particular deep and long-lasting conflict that had risen to the status of legend. During the course of my interviews, Jolene was consistently identified as "being able to get any man and any

2. Outside the mine, the extent to which women miners met men's standards of beauty was a source of concern to miners' wives. Ellen told a story about a miner's wife who "seen Darlene one day and said, 'Well, if all of them looks like Darlene, we don't have to worry.' Then she come to pick up her husband's check and Jolene was picking up hers and Jolene said 'Hi' and she said 'Hi.' And she said, 'Well, I thought you said all of them looks like Darlene!'" Darlene's appearance was within the confines of an "allowable image," while Jolene's was not (Fonow 2003, 77).

job she wants" because one led to the other. But a few women denied this. Ellen said, "They act like the reason she's got jobs is because of her body and her face, [but] it ain't that, honey. It's because she had enough sense to advance herself up and she's a real threat to some of the men." Kelly, who had known Jolene since she was twelve, said, "I think she gets a raw deal from everyone. I could not say enough good things about her. She works hard and has worked hard ever since she came there. She has never given anyone any reason not to like her." At the same time, Kelly admitted that she and other women at the mine simply might not know the whole story. Women who claimed to be "in the know" were quicker with their criticism. During our follow-up interview in 1995, Lynn told me of her suspicions regarding Jolene's long-standing relationship with an influential male miner.

> He seen to it that she got trained on this job and it was simply because at the time he was seeing her. A lot of men would have liked to have that job. It wasn't right, because that is a job that somebody with lots of seniority should have had. If it hadn't been for him, she probably never would have gotten a chance to be trained on it. So she went and asked to be trained on it and evidently he did too, because they let him do it.

Jolene had a particular nemesis in Carla. Although Jolene's alleged affair with the influential miner was fairly well known, Carla did her best to perpetuate it. "One time she had taken paint and written on a stopping [that I was] the [company's] number one whore," said Jolene. "And I mean, I hadn't done nothing. She really caused me some bad days." One of those "bad days" occurred when Carla played a malicious prank on Jolene. In Jolene's words:

> I came out [of the mine] to take a shower one night. They have four-inch plastic pipe they cut in two to make a tray in the bath-house to lay your stuff on. I always used this one particular shower, so I come in and lay my shampoo and stuff up here. She had taken belt grease and had put it all on the underside of the tray. I got it all over my stuff. I just waited until she was gone and I took what was left of it and I put it on the inside of her locker. I knew, there was no doubt in my mind. I know who

> did it and she never even mentioned it being on there to anybody. It made me know she did it.

When I saw Carla again in 1995, she was no longer working at the mine. This was much to Jolene's relief, because she didn't "have any trouble really with anybody" else.

> The biggest stress to me at this mine was her. I hate to let somebody get to me like that, but they can when you got other problems on [your] mind and you have to deal with little catty stuff like that. If I saw the woman go through a rock crusher at the feeder I would have been tempted to back it up. That's how much I hated her. That's how much trouble she caused me. I hate to let somebody get me in that position. She just tormented me mentally.

But more important than the ill will between Carla and Jolene was the effect it had on the rest of the women. Renee remarked, "We laugh about it, but it's sad." Most women recognized the flaws and foibles of both women, saying that they got along with both of them because, after all, both of them were good workers. They concluded that the true cause of the rift was the men's attention to Jolene rather than Carla's gross exaggerations of Jolene's infidelity. But, to make matters worse, at least one of the other women deliberately fanned the flames. According to Carla:

> I had one man describe to me exactly where my birthmark was, and it's out of sight, you know what I mean? And I said, "How in the world do you know where my birthmark is?" "Well, so-and-so [another woman miner] described your body. We asked whose body looked the best? Yours or Jolene's. And she said yours did." But we had women that was going in there and talking about us like that. Anyway, they rated me as having the best body up there. The women would go tell the men that!

Jolene corroborated the story, saying "We have one in there that told the men the location of every mole and everything on everybody. It was bad." Thus, the men's definition of externally defined beauty, along with the help of some women who adopted it themselves, had furthered the competition between them and threatened to erode their solidarity.

Previous research about women miners by Yount (1986) identified two ideal types of women based on a distinction between "working women" and "flirts." My interview data suggest that these categories represent two ends of a continuum.[3] The women at the mine could be divided into three camps. First, there were those who relied exclusively on their skills, knowledge, and work performance to gain advantages, such as a promotion. Whether they were deemed attractive or not, some women chose to accentuate the work role of miner during their day-to-day interactions underground. Second, there were those who chose to rely primarily on their physical attractiveness or sexual favors, or both, usually for easier work assignments. They opted to use interactional styles with the men that accentuated their sexuality instead. Third, there were a few women who used both strategies. They not only demonstrated the requisite skills and hard work for advancement, they also "hedged their bets," using their physical attractiveness to get work-related favors.

Any woman at the mine, regardless of her work capabilities, who was believed to exchange sexual favors for an easier job or for advancement was resented by most women and men alike. But most women agreed that those women received only a temporary gain. As Carla said, "They would probably get a better job until they all had her, and then she'd go back down." Those women who preferred to work hard as a matter of personal pride looked down on women who flirted or slept their way to a favored position. Lynn said:

> After I seen how some of the women was, I was bound and determined that the men would never talk about me like that. I was single, too, and I could've done the same thing they done. I just couldn't handle working if I knowed the men talk about me like they do some of those women. There are some that still does it. There's just very few. Maybe two or three of them of all that we have there might be like that. Well, there's one that she

3. Specifically, the two ideal types of women miners established by Yount (1986) are the "competent asexual tomboy" and the "incompetent feminine flirt." My data indicated that women's work-related competence and the extent to which they accentuated their femininity as they worked were not that closely related. Thus, I propose using these two extreme ideal types as opposite ends of a continuum.

> don't have to flirt now because her husband's a boss. [*Chuckles softly.*]

Unfortunately, women who traded sex for lighter duties reinforced the image of woman miner as gold-digging whore and made it harder for other women to resist that image. Sally told me:

> The ones that don't have affairs have to work harder, and the other ones walk around grinning. So there's friction there. And then some of these girls that I have worked with myself would have affairs with these bosses, and another woman would come on the section. He would send that woman he's having the affair with [and] he would take the woman who come on the section and send her with her, and both of them would be working at the same job hardly doing nothing. So that's the way a lot of the bosses done. He would send the women together rather than send the men together on an easy job.

Some foremen took an equally dim view of these women because they created unnecessary divisions among miners. Lynn said most foremen preferred not to deal with flirtatious women on their crew.

> The few that I've talked about that's got the flirty attitude, the bosses just as soon not to have them because they know what they're doing, and they know that the men and the women know what they're doing, and if they show preference, it's going to come out. They just as soon not have them on their section as to have to put up with the hassle of it. Eventually they are against that kind anyway. It didn't get them nowhere. The ones that are like that have been that way. They're still not doing nothing, and nobody thinks much of them.

To the extent that men continued to view women categorically, each woman at the mine represented her gender. Flirtatious women who reinforced other women's objectification compromised, knowingly or not, the political power that all women could have used to assert their collective legitimacy in the workplace. Therefore, the women had to resist their objectification and sexualization individually.

Citing the issues that have divided them over the years, several women miners complained about their lack of political solidarity. Those who complained the most were the ones who readily recognized the inequities inherent in a male-dominated work setting and the resulting discriminatory practices. They tended to more assertive than the rest and more interested in the challenge of their jobs as a means of gaining self-esteem. These women were more cognizant than their peers of how unity could change the lot of all the women at the mine. As Ellen said:

> If the women would stick together, we would be recognized as different people, and it's not that we want to be, but we are. That's what the problem is, ain't it? The men knows that, too. They're all afraid of stepping on somebody's toes. I said, "There ain't no way you going to get worser off. It don't matter what you do or say." I believe if they would stick together, there would be new meaning up there. They would recognize us. But right now I don't know what they think of us.

One veteran male miner said he wished there were more solidarity among the women. "But there's none whatsoever," he said. "I mean, it could be a good thing if there were more solidarity between them on a local level, because they could all get together and have some input to what goes on at the mine. If someone was treated unfairly or if they wanted better working conditions, the women could do more things if they'd stick together. But they won't."

The women blamed their own lack of solidarity and collective resistance on their dwindling numbers and relative social isolation. Irene said, "They don't have a lot of women up there no more, which they should have to put some women in there." When there were more women, "they would try to get involved with other stuff that the men was doing." Ellen agreed but pointed out, "We're not strong enough, and that's the reason that not many of us stayed there. Because if we would have helped some of them women, somehow maybe some of them would have stayed." During our follow-up interview in 1995, she added, "I've always said, what we've needed to do was actually go and do things together and travel and get to know each other outside of the mines, you know?" Not only were the women at the mine somewhat socially isolated from one another, very few of them knew women who worked at other mines in the area.

They also said that finding the time to strengthen their bonds was much more difficult for them than it was for the men. Most of the women had families, which left little time for socializing or having serious, uninterrupted group discussions. "I don't socialize," said Carolyn. "I like the people, but I don't have time to socialize. You work six days a week, ten hours a day. There's always women who socialize a little, but the men socialize a lot." The men "don't usually go home and run errands or wash clothes. Most of them have wives at home." The men I interviewed confirmed that they got together outside the mine much more often than the women did. Their social interaction deepened their bonds and strengthened their collective sense of masculinity, which also solidified the patriarchal control they had over women at home and at work (Gibson-Graham 1996).

Apart from occasional outings among small groups of women miners, the women's bathhouse was the only place where women could engage in "safe discourse" (Collins 1990; Evans and Boyte 1986). As I observed them between shifts, the women frequently and freely interacted there, trading stories from their own standpoints as women and establishing common bonds as women in a male-dominated workplace. Here they were uninhibited about sharing their subjugated knowledge and expressing their opinions and occasional outrage. Many of my data about the women's relationships with each other, their feelings about the union, and their work-related problems were collected there during shift changes.[4]

The women also shared strategies for resisting patriarchal control, particularly the power of controlling images. But, since most women's family obligations kept them from attending union meetings or informal gatherings, and since few of them worked together on the same section or shift, their free space was severely limited relative to the men's. Like women steel workers, they "did not have enough sustained contact with each other on the job to develop a communication network that could address gender-specific issues" (Fonow 2003, 94). Their discourse in the free space of the bathhouse was their primary opportunity for collective empowerment.

When I interviewed these women again in 1995–96, most of them spoke readily of their lasting friendships and the mutual respect they

4. During one of my many visits to the women's bathhouse, I listened to Darlene and another woman exchanging information about the ways they successfully solicited sex from their husbands. This backstage behavior was clearly not meant for front-stage, cross-sex consumption (Goffman 1959).

had developed for one another over the years. They stressed the need to maintain good relations with one another, particularly necessary for getting work done underground. Most of them tended to recognize and praise one another for their good work records and reputations, and told me, "Sadie's an excellent worker, your model coal miner." "Ellen has a tremendous work record." "Darlene was a brute. She could outdo some of the men." Those women who had the best work reputations were also generally supportive of women miners whose work performances were weaker. Despite their lesser endurance or skills, they said, those women "gave it everything they got." As Kelly commented, "We could say so-and-so's lazy, but they may be working up to their potential. They may not be working up to mine, but they may be working up to theirs." In addition to a good work reputation, a woman's pioneer status affected her informal standing among the women. Most of them looked to Sadie as a benign matriarch. "She's like a mother to us all," remarked Kelly, one of the youngest women miners. "I have a lot of respect for her."

These solid work relationships formed the basis for some lasting friendships. Although the men saw no evidence of this, all the women I interviewed talked about their friendships with each other. During the latter part of my field research in 1990, as a matter of protocol, I tried to keep from disclosing the identities of the women I had already interviewed. During a phone conversation with Kelly, I learned that my methodological resolve was a real-world absurdity when she blurted, "We discuss things amongst ourselves, you know!" She let me know that the women at the mine were much more than a mere collection of coal-mining women. During our 1995 interview, Ellen told me, "There's not much that divides [us] any more." The women said they depended on one another for various kinds of support because, better than any of the men, they understood each other's problems. As Carolyn said:

> In the bathhouse, they talk to each other and they sit beside each other. Somebody's having an operation and we're all worried about her. That's the way the women are. The women are concerned about each other like that. If a woman has cancer, they can really understand that, because they're a woman. A man can't understand that like another woman can. They have babies and something is wrong with their babies, I mean, they can really understand that. They can sympathize because they have kids.

Kelly added, "I think we stand up for each other more now than we ever have because we've all got the same little problems. We all understand each other better because at the age we are now, we're all having health problems. I think that made us understand we got to hang together." In Carolyn's words, "All of the women are friendly to each other and in my opinion have a lot of compassion for the men as a people thing. These people are really good people. Sometimes we don't understand [when] they come to work and they bring a chip on their shoulder. They're just under stress and you feel bad for them. I guess that's just generally people." In an even more compelling statement about the miners with whom she worked, Ellen said:

> I don't see the bad side of most of them because everybody's got their bad side, ain't they? But they're just hardworking poor people and I basically feel sorry for them, because you see them come in there and they might have a mashed hand or cut hand or dirt under their eyes where they didn't wash their face good or a button off their clothes. You look at their sandwiches in their lunch bucket sometimes, and they don't have much. It seems that they should have a better life, all of them, because they work too hard not to.

In sum, although the women at the mine had done little to make their political presence known, their cohesion and contributions were enormous relative to their numbers, as they became socialized to the men's world of underground coal mining.

EPILOGUE

When asked how mining had changed them, the women in my study talked about working hard, having to be tough with men at times, and adopting more subtle strategies for coping with them at other times. Gaining the men's respect was more important to these women than anything else, whether this involved resisting or accommodating to the masculine-identified world of coal mining.

Resistance came more easily to some women than to others. Sally said that working underground made the women she knew tougher. "It might have made them tougher as a human being, tougher as a person, because there's things that they do that they probably thought they never could do before," she said. "I was raised rough, so I'd just about tackle anything myself. It didn't really bother me." For others the transition was more difficult. In 1990, Ellen said that when she first started working underground she was "scared to death" of the men. "For the past ten years I felt like I was the underdog, that I shouldn't be stepping on their toes. I wouldn't talk back, Lord have mercy. It wouldn't have mattered what a man said to me, I'd just take it, really." Five years later she said, "I was afraid of a lot of stuff and I wouldn't speak out. I was really taught to sit in the back of the bus and keep your mouth shut. That's how I felt I was taught. But I don't keep my mouth shut no more. I just say what I think, which most times is pretty good." Amused by her own transformation, she concluded, "Yeah, I've changed a lot." Two other women agreed that none of the women working at the mine became upset over conflicts with men any more. Michelle said, "I think all of them are all right, they been in there long enough. You be shoving me, I be shoving you. I think the average woman over there been in the coal mines seventeen years. I think now they are speaking up more over time." Having gained the confidence to do their jobs, most women said they had developed greater respect for themselves and were more aware of the issues that affect working women in general.

To the extent that some men continued to view them in terms of their attractiveness, the women still felt some pressure for "doing gender" (West and Zimmerman 1987), but the stigma of doing a "man's job" had faded over time. They were now seen, and saw themselves, more as miners than as women miners. According to one of the veteran male miners:

> I saw [*names four women miners*] come here as basically young ladies [who] have gradually gone through changes. I saw it like when you could walk out here when more of the ladies were attractive. There were several of them here at one time. Super attractive. And, you know, there was just people around them all the time. But you're seeing that fade away and I guess it's that they are either losing their ability to be attractive or the men are losing their grip!

While some women never felt any pressure to be feminine, others believed that over time the importance of being feminine had receded. They were there to work, just like the men, and they regarded femininity as a poor way to measure them as miners. Irene said pointedly, "I didn't feel no different. I was there, they were there, making the same pay." In 1990, Ellen told me, "I guess if you're going to be down there, you get more and more like you're a man. It takes a lot out of you, like dresses and stuff. You wouldn't hardly see any woman in a dress outside the mines anywhere. There's nothing delicate about it, it just changes us all over." When I asked if that bothered her, she replied, "Yeah, it does. It really does, because I think, women are women, they have to be. There's no way a woman can be a man or a man can be a woman, there's differences." Five years later, she said, "When I go down there, I don't feel like I'm a woman or a man or nothing. I've always said that I feel like I'm the coal miner, I'm the person who has to do the job. When I come back out there, I like to feel that I'm a woman again." She concluded reflectively, "Well, it might just be how everybody's changing in the world, not just us, you know?" Over the years, the women had resolved the contradiction of their doing a man's job; their accommodation to work became a form of resistance to traditional gender roles.

But the women also discovered that there were limits to challenging men's beliefs about their superiority and power in the workplace. Referring to the persistence of perceived gender differences, Sadie remarked, "It's never going to erase. They understand that we are there and we are women." Indeed, during our 1995 interview, Lynn said there were still some men who refused to accept them. "I'd say no more than 10 percent. [But] they don't say too much. There are some men up there that won't hardly talk to you. I think most of them like that, their wives are jealous." She added that some men still resent having to help women on the job. "A boss can tell them to go help you and they resent it. They really feel that we really didn't come to work, that we just came to get the job and the money." These same men took a dim view of women who had advanced to more responsible, higher-paying jobs underground. Ellen agreed that some men continued to "think that the women shouldn't be there and they shouldn't be making as much money as they are. That is basically how they feel, and why should a woman make top pay on a job a man should have? That's a man's job." Carolyn suspected that men who voiced their opposition to women miners "just try to goad [the women,] saying, 'You should be at home.'" Her response to this attitude was:

> Yeah, I want to go home. I'd like to be at home at my kitchen table. I'd like to stay home more and take care of my kids until they grow up. I would like to be home letting a man take care of me. They tell [the women] they don't belong here. Go home and take care of your babies. Well, I had my babies. I'd love to go home, sit at home. Some of them say it just to get you going. I just agree with them.

Paradoxically, her acquiescence amounted to a form of individual resistance to men's oppressive beliefs, because the men's remarks had no effect on her definition of self as a competent coal miner.

In their struggle to become integrated underground, the women discussed using certain strategies of accommodation during their everyday interactions with men. "Two-thirds of the time I hate men," Ellen said with a hearty guffaw. "Ain't that awful? I just see how stupid they are. But I try to be as nice and as fair and do my job as best as I can. And when I go in there I don't crack no dirty jokes. I don't say nothing dirty. But I do think of some stuff." She stressed the importance of maintaining good relations, even if on the men's terms, because "to me, they're about the only family I've got anymore." She said that if she thought something was wrong, "I would ask them had I done something to them and they'd say no. I'm mad because of so-and-so or because of this and that. But I always ask them, because I think they're not mad at me, they're just mad at something else and I'm there to take their spite out on."

Taking a more proactive approach, Kelly admitted using men's egos to negotiate her place among them because "that's a man's weak point." She thought the best way to get along with both men and women underground was to "to keep my nose on my face. You don't want to get told off because you say something totally innocent about someone and at the same time that gets back to them." Both women also pointed out the advantages of remaining aloof and letting the men take their silence as tacit agreement, although it did not signal their submission. Ellen said, "That may not make me feel better, [but] I think, well, that works pretty good." While Kelly said, rather philosophically, "You always have one or two buttholes in every walk of life. But those people you stay away from. One thing I've learned to do, you can't argue with ignorance even though you are right. What I do is, I will let them assume they are right. I agree with everything you say and go on. It keeps the argument down." The women's overt acquiescence, assumed to be "natural" for them, was actually a form of their covert resistance.

From the time they first set foot underground, the women understood that coal mining had always been, and to some degree would remain, a male domain. "I knew when I started I was going into a man's world, and men have their ways that they done a lot of things before we got there that they had to change their entire ways," said Kelly. "When the first women went into the mines it was hard for a man to change his ways. They expected a lot out of them." Previous research has shown that when women have entered male-dominated occupations, men demonstrate their own accommodative patterns toward them (Enarson 1984; Swerdlow 1989). But because women challenge men's status in the workplace, men's beliefs and behaviors about them allow them to reassert their superiority while simultaneously accepting women as co-workers. Both women and men said that a majority of the men had come to accept the women who worked in the mine. "Attitudes have changed," said Carolyn. "I think most of them have changed [among] people that

I know. A lot of the men know that the women can work hard and that they are there to work." "I think we gained a lot of their respect," said Renee. "They're not as quick to judge a woman about what she can't do. Everybody has changed in one way or another, the guys' attitude and the bosses' attitude." Michelle added, "Why call it a man's occupation? I think we made a big change in the industry, in bosses' and men's attitudes, because we stayed."

During our brief conversation, a miner and former union president remarked that over time he had favorably "changed his opinion about the women miners." Although somewhat more cautious in his assessment of them, another male miner said, "There's a place for women in the coal mines. If they're cut out to be that way, if that's the way they are. Some folks look better in an evening gown than they do in a pair of Levis. They're not going to make it in the coal mines. But some women can make it in the coal mines. They can, and they make good money, good hospitalization."

The men's acceptance of women underground was also based on their recognition of the women's practical needs and their strengths and skills as individual workers. When it came to earning a decent wage, Kelly remarked that "the men have accepted us as equal partners and it's not like what's good for me isn't good for you." Michelle said she thought that "75 percent of the guys up there believe that we have a right to work, too. You have a right to support your family, too." Likewise, one of the men pointed out that the women he had worked with at the mine "have to work because [they] are divorcees. They have no other income." Judging the women on their merits as workers, one miner said he believed that most of the women were just as capable as many of the men. Another said, "I've worked with several of them underground. They work hard and are concerned about how they perform their job. I think they are very qualified." Finally, the women themselves also believed they had gained recognition as individuals. "Most women up there now do try," said Renee. "They're trying harder now because they've established their position." Over time, Irene remarked, "They knew who they could depend on and who they couldn't depend on. Now, if they sent me to do a job, I tried my best to get it done."

In sum, although to some degree they were still subject to the men's definition of them as workers, the women's tenacity and hard work had legitimated them as coal miners. By the mid- 1980s, said the superintendent, "The novelty of having women underground wore off and you found out who the keepers were." Both women and men miners had matured, he said. "We've all grown up and it has brought about the integration of the sexes. Complaints about women are passé." "Now," said one older male miner, "they're just part of the operation." Looking back over her years at the mine, Ellen commented on the adjustment process, saying, "Now, basically, the men I work with got a lot of respect. They're family men and I guess we've growed onto each other we been there so long. I guess they got kind of used to us, and they don't pay attention to us." Kelly added an interesting insight. Since women had started working underground, she said, the "men respect themselves more. They are more cautious and they would not lower themselves to do anything in front of us. They respect us, too." She concluded, "When I started at that mine I was nineteen. And when I started, the average age was twenty-three. So we were kids going underground on a Saturday night. I look at all of us now and we're all gray. There's some gray in all of our hair. We have all grown up together. I think that's why things are better."

FACING THE FUTURE

This book has focused on the conditional and sometimes uneasy alliances women miners formed with men underground. Together they now face an uncertain, if not downright bleak, future, as do most underground coal miners working in the mountains of central Appalachia. Decades of mining have taken both a physical and a psychological toll on the miners, women and men alike. Physically, they have suffered injuries, some of them requiring surgery, that continue to plague them. Day in and day out, miners spend most of their time working harder than ever before, the most dependable among them being "rewarded" with more work. Psychologically, a sort of mental fatigue has set in. Kelly admitted:

> Every day we discuss how our day went, what we had to do. Had a hard day. It's just a natural normal routine. We all have a bad habit of saying, whether it's true or not, "I'm tired. I'm busted, I'm killed." I'm telling you, I say it in the mornings before I even get my clothes on. Oh, I'm dying. It's mostly physically

> tiring, [but,] like I said, there is so much mental stress on us, like, is our job going to be here tomorrow?

Since the early 1980s, coal-mining jobs have been rapidly disappearing in the coal-dependent areas of central Appalachia, and miners have every reason to be worried. Advances in deep-mining technology and competition from surface mining, mountaintop removal in particular, have made underground miners increasingly obsolete. In 2000, only fifteen thousand underground miners were working in West Virginia's mines, and a 23 percent decline was projected by 2010 in the coal-rich counties of the southern Appalachians (Lipton 2002). Thus, despite record production, employment is at its lowest point ever. "If [the mine] lasts ten more years it will be a lifetime, because the plow system's mining more coal than we have ever mined," Kelly told me. Voicing the miners' collective concern, she said, "Nothing's going the way we want it to go. There's always a threat of a layoff. The biggest threat right now is, how long do we have? See, you are always threatened, and most of us are at the age where we will all be reaching retirement at the same time. So we know that [the coal company] is going to put it to us because of how they have done retirees already." A union official echoed Kelly's sentiments, saying, "The way things are now, I'm not sure about a lot of the people that are here. There are going to be a lot of people out there hurting. You know, if they don't make it through the system here to where they can get retirement, it's going to be very difficult in their older years." Voicing a concern of his own, he added, "It's really hard to talk to some of the people, because people just don't want to accept reality, [but] sooner or later it happens." Kelly continued, "We have no outlook. It's today. I will have to work until I drop dead. I've been in the mine sixteen years. There's no hope for me. You cannot live on a twenty-year retirement. Work until I die. That's all I can do. That's my philosophy. I worry about today. I worry about a rainy day tomorrow."

The superintendent declared that "daily we face up to being a dinosaur. This organization is having to face its own mortality, and in the global structure there just aren't enough jobs." In his own way, he was saying that the miners at the mine will eventually have to face losing their jobs forever. Women are particularly at risk because they were among the last miners hired and "because there's more women that just never got involved in doing all the work," said Lynn. "If [the company] gets to where they start contracting out jobs [like] all the firebossing, all their belts, all their masons. We've got women that can't run a piece of equipment in the face. It would come down to your face people was all that's left at the mines. The majority of [the women] would be gone."

Indeed, some women retirees already live with that regret. "I really like the mine. I like everything in it. I can't explain it," said Irene. "I wished I could have twenty more years. If I could go back in, I would do a lot of things different. I would get on the machinery and things like that instead of building stoppings and things that I did do." If returning to work was out of the question, the prospect of a daughter or son getting a job underground was even less feasible. "If I could get my son or daughter a job here tomorrow, I would," said Jolene. "I mean, you know there's always the danger, but there's that out in the hollow. It's a good job here, good pay, good benefits. I would encourage them, [but] there's just no jobs here. I don't think they'll hire red caps again." Recent headlines have proclaimed that recruits eager to replace retiring miners have begun lining up for training, but there is no news about efforts to recruit women (Sheehan 2004). "I would like to see more women in the mines," a male miner said to me, "but I don't think we will. I don't believe you'll see any more women coal miners."

Several years have passed since I interviewed anyone working at the mine. Only recently I learned that during the sweltering summer of 2003, the miners' worst fears came true. In an ominous twist of fate, a lightning strike disabled operations at the mine, shutting it down indefinitely and leaving miners' lives in limbo.

Patriarchal ideology assigns a woman's place among men, capitalism assigns women's and men's material places as workers in the work hierarchy, and colonization assigns Appalachians a place as members of a region renowned for the exploitation of human and material resources and grinding structural poverty. Although their collective identity, solidarity, and potential for collective resistance will never be fully realized, the experiences of women miners working in the coalfields of southern West Virginia represent the existential intersection of these virulent social and economic forces. In this context, women miners attempted to fend off the controlling images of them based on gender, race, class, and region. To some degree, they succeeded. Their awareness of their own subjugation and their development of a dual consciousness formed the basis for individual resistance as they negotiated their individual identities. Patriarchy prescribes that a given individual is either masculine or feminine, that each gender is essentially distinct and dialectically opposed to the other (Parker 1992; West and Zimmerman 1987). Venturing beyond their gendered socialization experiences, these women deconstructed gender. From the margins of their gendered positions, these working-class Appalachian women became agents of social change that has formed the bedrock of their legacy.

APPENDIX: FIELDWORK AND PROFILES OF THE STUDY

This appendix spells out how the study was done, detailing the use of different qualitative techniques, my experiences in the field, and how I handled different issues that arose during my fieldwork. It also presents a profile of the mine, the classification of jobs in an underground coal mine, and the sample of women miners I interviewed.

FIELDWORK TECHNIQUES AND REFLEXIVITY

"Feminist standpoint epistemology" necessitates the use of techniques like those featured in this book.[1] Because what women miners know

1. Feminist standpoint theory and the realities of women's working lives are linked by feminist standpoint epistemology. Standpoint epistemology posits that because women are among the less powerful members of society, they belong to a separate culture (Collins 1990; Harding 1986; Hartsock 1983; Jaggar 1983; Nielsen 1990; Smith 1987, 1990). As a result of their subordinate status, they develop their own understanding of the world, separate from men's understanding. Feminist standpoint epistemology involves "a level of consciousness and awareness about one's social location and this location's relation to one's lived experience" (Nielsen 1990, 24). Moreover, it is incumbent upon women to be attentive to men's perspective in addition to their own because their very survival in a man's world depends on it. Because of men's dominant status, the reverse does not hold true. Thus, relative to women's viewpoints, men's viewpoints are not only partial but are contradictory as well.

Feminist standpoint epistemology places women at the center of study that begins with women's experiences but does not necessarily end there (Nielsen 1990). This is not to imply that women's knowledge is superior to men's but rather that, owing to the general neglect of women's experiences by more conventional researchers operating within more dominant paradigms, it is necessary to focus on it. Moreover, as with most empirical approaches, it is understood that focusing on one type of phenomenon during such an investigation means forfeiting emphasis on another (Vidich and Lyman 1994). Because my study is informed by a feminist standpoint, it is inclusive rather than exclusive of men, unlike androcentric approaches that ignore women altogether.

about their world, particularly about their workplace and the men with whom they work, constitutes the central focus of this investigation, three techniques were used: in-depth semistructured interviews, observation, and archival research (or document study). The use of multiple sources allowed me to use same-method (i.e., interviews) and multimethod (i.e., interviews, observation, and document study) triangulation. Comparisons of informants' accounts and information obtained from different methods strengthened the validity and reliability of the data and their subsequent analyses (Lincoln and Guba 1985). These techniques also brought me close to the data and allowed me to develop explanations directly from them for verification, modification, or extension of socialist feminist theory (Fielding and Fielding 1986; Lincoln and Guba 1985).

Knowing the methods I would use to collect data gave me some of the direction I needed, but a lot of preparation goes into fieldwork. Not knowing much about coal mining, I began poring over books about mining technology and the industry's labor history, and learning both the technical language and slang terms used by coal miners.[2] I obtained copies of the 1988 and 1993 contracts between the United Mine Workers of America and the Bituminous Coal Operators of America in order to become familiar with particular clauses about absentee policy, promotion, realignments, and employment termination. Over a span of ten years I attended four conferences of women coal miners—1988, 1990, 1995, and 1998—sponsored by the Coal Employment Project, which

Moreover, unlike the more dominant epistemologies, feminist standpoint epistemology rejects the notion of objectivity and maintains that knowledge can only be known intersubjectively. That "intersubjectivity of meaning takes the form of a dialogue from which knowledge is an unpredictable emergent rather than a controlled outcome" (Westkott 1990, 62). In Seitz's (1995) landmark work on the lives and work experiences of contemporary Appalachian women, she quotes Paulo Freire (1972, 64) that "authentic thinking, thinking that is concerned about reality, does not take place in ivory tower isolation, but only in communication." Furthermore, as Denzin (1989) reminds us, research itself affords the researcher a mode for investigating the actions of others. Naturalistic research and qualitative methods bring into sharp focus the role of the researcher as well as the situations of those being researched. Therefore, two equally important components of any research of this kind are the researched lives under study and the research as conducted by another socially situated individual. The corresponding methodology requires a certain reflexivity as the researcher conducts herself in the field.

2. The following sources were invaluable to my preparation for fieldwork: Keith Dix's *Work Relations in the Coal Industry* (1977) and *What's a Coal Miner to Do? The Mechanization of Coal Mining* (1988), Charles R. Perry's *Collective Bargaining and the Decline of the United Mine Workers of America* (1984), William P. Rogers's *The Coal Primer: A Handbook on Coal for the Non-coal Person* (1978), and Alexander M. Thompson's *Technology, Labor, and Industrial Structure of the U.S. Coal Industry: An Historical Perspective* (1979).

gave me the opportunity to talk to women miners about their on-the-job experiences and concerns before I went into the field.

These experiences were invaluable in preparing me for fieldwork. At the first conference, in 1988, I spoke at length with Bernice Dombroski, who was working at the Rushton mine in central Pennsylvania during and after Marilyn McCusker's tragic death. Several weeks before I went into the field in 1990, we met at her home, where she helped me test my questions, making suggestions about how to ask them and what additional questions to ask. At the 1990 conference, I spoke with former miner and then UMWA photographer and journalist Marat Moore, who suggested two potential study sites based on the mines' size and greater numbers of women miners. When preparing to contact these two companies, I sought the advice of an established mining engineering professor at Penn State, Stanley Suboleski, who helped me draft letters to the companies and let me use his name as a reference. The first coal company I contacted was apparently still smarting from several lawsuits filed by women miners over sexual harassment, because its human resources director flatly refused my request. The second company I approached agreed to cooperate with me on the condition that I preserve the anonymity of the company, the mine, and its workers, and we made the necessary arrangements for my visit.

Because a researcher's preconceptions can color her interaction with informants, I treated my own as the basis for comparison with what I experienced in the field. (See Nielsen 1990 for a discussion of this process.) For example, I attempted to avoid the trap of essentializing the women I encountered by being ever mindful of the diversity of viewpoints among them but also recognizing the singularity with which they spoke to me. I was also wary of romanticizing their lives and occupation. While I attempted to locate *myself* and my own lived experiences in relation to those of my informants, I realized that their sense of being a part of, but still separate from, the masculine-identified world of mining paralleled my own awareness that I too was immersed in a world to which I did not truly belong. In this sense, as Westkott (1990, 61) has observed, "women studying women reveal the complex way in which women as objects of knowledge reflect back upon women as subjects of knowledge. Knowledge of the other and knowledge of the self are mutually informing, because self and others share a common condition of being women."

I also dressed and conducted myself in such a way as to minimize the class differences between myself and my informants, in an attempt to develop a rapport between us. Most of the women were about my age and dressed as I did, in a flannel shirt, jeans, work boots, and a faded army jacket. As the interviews got under way, I established myself in the role of "student" eager to "become educated" about mining by my informants, whom I called "my teachers." I likened my own advanced degree to their getting their mining papers, the certification they needed to advance beyond mining trainee. Erring on the side of caution, I avoided using such "red flag" terms as "harassment" or "feminism." Instead, I asked about "difficulties with men co-workers," referred to "equality," and used such phrases as "equal work for equal pay."

But I soon found that there were other differences I also needed to minimize. When soliciting or initiating an interview, I introduced myself as a Penn State graduate student in 1990 and as a university professor in 1995–96. Even so, I was sometimes asked with suspicion if I was with either the company or the union. During the 1990 fieldwork interviews in particular, I was occasionally identified as being "from up north." I told my informants that although I was at that time attending Penn State, I had grown up in Virginia, thus giving myself more "southern" identification. My first few interviews were crucial in establishing my legitimacy as a researcher and thus in gaining the women's trust. As several of the women I interviewed told me later, the first few women I had interviewed had assured the others that their interviews had not resulted in any negative consequences, thus putting them at ease. In addition to the fourteen women who agreed to in-depth interviews, seven other women I spoke with agreed only to informal conversations, fearing reprisals from the company. Typically, they responded to my solicitation, "Sorry, but I need this job." There were at least two more women who flatly refused any contact with me. I later discovered that one of them, who was married to a boss, had been ostracized from the group for allegedly informing company personnel about union activities. Asking her for an interview in front of other women actually provided some of them with another opportunity to harass her.

Understandably, most of the women I interviewed were as curious about me as I was about them. I was careful to limit my own self-disclosure to avoid losing my "trusted stranger" status in the field (Warren 1988). Anthropologist Ann Kingslover (1992, 103) has advised that "negotiation of an ambiguous identity . . . can be used to resist constructions of power relations that rely on fixed notions of identity." At the same time, the women's comments reflected their level of acceptance of me. For example, after some discussion about the rigors of mining, Kelly and Alice commented on my "athletic build" and exclaimed that with the required training, "You could be on our crew!"

Ellen told me, "You could probably do it pretty good, [but] you know the reason [the company] wouldn't want you is because you've got enough sense to advance yourself. You'd be pushing all the time to learn more and more, like I am now." This brief comment said much about her perception of the relationships among women and between women and men at the mine. Only a few women shared a common ambition, which made them undesirable because it challenged men's power and status. In our case, we were united by our mutual ambition, which represented our opposition to male privilege. Carolyn asked me if I was going to go underground. I told her I had arranged a tour with the mine superintendent and asked her how I could think of not going into the mine for a mere eight hours, when they had been working there for a decade. Smiling and nodding her approval, she said "You'll understand when you go underground," as if doing so was to be my own "rite of passage" into the world of mining.

Finally, one day during my fieldwork, I traveled to a neighboring town to buy some supplies and take a break from my routine. Upon my return to the women's bathhouse the next afternoon, I was greeted at the doorway by Alice, who assumed a posture of impatience, with one hand on her hip as she spoke. She, Kelly, and Carla expressed concern that I had either run into some trouble or left without saying goodbye. This told me that I had become an accepted fixture in the women's bathhouse.

While my similarity in age to the women I encountered helped me negotiate my identity among them, it may have worked to my disadvantage when interviewing men. Given that managerial personnel had repeatedly referred to me as "the lady here to talk to our lady miners," I soon discovered that the men believed that I was only interested in "women's problems" and not in their perspectives and experiences.[3] Moreover, when I asked the women about the men they worked with, they were generally forthcoming with both negative and positive comments and stories. However, when I spoke of interviewing the men, the

3. Although the term is meant to be polite and respectful, previous research on the labeling of women posits that the term "lady" implies not only high standards of conduct for the female so designated but also assumes the maintenance of her protected status and therefore of her delicate and impotent nature (Martin 1980).

women expressed anxiety about being identified as the source of their names. Even though I assured them that as a rule I did not make such identifications, they still voiced concern. Ellen asked, "You're not *really* going to talk to them, are you?" I took their apprehension seriously because I felt an ethical obligation to avoid causing any of them undue stress about the conditions of their employment. (For a discussion of these issues and how to cope with them, see Punch 1986.) Thus, I limited my contact with men miners to those with whom I had my own incidental contact and those who were visible because of their positions in the local union.

During the course of my fieldwork, I described the study to my informants and explained their rights as voluntary participants. I have chosen pseudonyms for each of the women quoted in this volume, in keeping with my promise of confidentiality. Sometimes informants would ask if I had interviewed another informant. I faithfully replied that I could not tell them but that I would keep their confidence as well. Although this often became socially awkward, especially when informants talked among themselves, I believe it minimized, if not eliminated, some problems that my presence could have caused. Even so, Carla told me that during my visit to the mine in 1990, the company posted three trainee jobs.

> Since you come up to talk to us in the bathhouse that day and set up these appointments to come and talk to us, they have put up three maintenance trainee jobs and this is something that they haven't done in maybe eight, nine, maybe ten years. Never put up no trainee jobs, period. An official from the union, we ran into him and I explained to him about you coming to talk to us. He said, "Maybe that's why, they know you all are going to say something."

During our follow-up interview in 1995, Carla also told me that shortly after I left the study site in 1990, management had questioned her about what she had said to me during her interview. Similarly, the day I left the site in 1990, I stopped by the mine office to tell management that I was leaving. This had become my habit, because each time I visited the mine, I was required to sign myself in and out. On this final occasion, I spoke with the secretary, with whom I had previous contact. She questioned me about which women I had interviewed. Obviously this raised serious ethical issues, and I politely declined to reveal the identities of my informants.

During our follow-up interviews in 1995, Renee and Ellen told me that I had aroused the suspicion of both the company and the men. A boss was worried about the questions I might ask about sexual harassment. Renee said, chuckling, "When he found out I was going to have this interview, my boss said, 'Come here.' I said, 'What?' 'Have you been sexually harassed by any of these people around here? By any of these bosses?' I said, '*What?*'" I asked her if the boss seemed worried. "He was worried," she said. "You can throw a big lawsuit and stuff." Ellen told me that my presence concerned local union officials as well. She said with a wide grin, "Well, I'll tell you something. They told me at work, they're scared to death of you." I asked why. "They probably think you, well, uh . . ." and she laughed. "That I'm a spy?" I asked, to which she responded, "Yeah. For [the company]."

These stories revealed the suspicion I provoked among some of the company men, but also the legitimation of my role as researcher among the women at the mine. My first interview with the superintendent was equally revealing. Generally well regarded by most miners, he started his career as a miner and had worked his way up the organizational ladder. He was an articulate and well-educated man who held a master's degree in history from a prominent university located just outside the region. He said he was particularly interested in speaking with me because he wanted me to get the "right" impression of the company.

Several other encounters I had with men at the mine increased my sensitivity to the gendered power dynamics that the women miners faced in their everyday lives at work and in the local community. The day I was to tour the mine, I met a mining engineer at the equipment warehouse and discussed what sections of the mine I wanted to visit. In particular, I asked to go to the face (coal seam) and watch the operation of the plow or longwall system. As I began to dress for the tour, donning steel-toed boots, a hard hat and miner's cap light, a belt with an oxygen self-rescuer, and a rain suit worn by miners who worked on that section, a miner who was standing nearby asked me, "You're going to the face? You *want* to go up to the plow?" I said yes, to which he shook his head and asked why. "Well, because that's where folks are working," I replied. For the next ten minutes or so, the same miner kept popping out from behind shelves of equipment, shaking his head and murmuring, "She wants to go see the plow!"

A few moments later he said, "It's dirty up there, you know that?" "Yes," I replied. "It's cold down there," he said. "Uh huh," I responded. "You'll get hungry," he said, "bring a candy bar?" I nodded. Shortly thereafter, the engineer and I went out to the portabus and rode down the track into the darkness of the mine. I concluded that if what you find underground is symbolically identified as "masculine" (i.e., steel castings and timbers, construction-type machinery, and heavy equipment), then the plow section, where coal is cut from the face, was the center of coal-mining masculinity, and I was an intruder. (See Cockburn's 1985 discussion about the gender-typing of machinery.)

The second encounter, however, was far more threatening than the first because it could have resulted in the revocation of my privilege to visit the mine facilities, including the women's bathhouse, where I solicited interviews. One Saturday afternoon I drove around the community taking pictures. At one point I pulled off the road several hundred yards from the main gate to the mine to snap a picture of the tipple. A small silver truck traveling behind me slowed down long enough for the driver to get a good look at me. I ignored him but made a mental note of it. While I was taking pictures from another angle closer to the gate, the same truck passed me, going the other way. Again, I simply noted it. I put my camera away before entering the compound because I had been forewarned by Jolene that the company did not permit the taking of pictures on its premises. I checked in at the gate as usual, filled out the required daily visitor's pass documenting the time and purpose of my visit, and proceeded to the lamphouse to record the miners' employment information that was posted there. But my recording was interrupted by a security guard who strode into the room, dumped the contents of my briefcase on the bench where had I left it, and asked gruffly, "Where's your camera?" I explained that I had left it in my car, and I had not taken any pictures since entering the premises. He seemed satisfied that I was being truthful and was visibly embarrassed by his mistake. He apologized and helped me get my papers back in order. Then he offered to "make it up to me" by taking me on a tour of the surface facilities. Oddly, he never asked me what I was doing at the mine. I believe he already knew. The first place we toured was the now deserted men's bathhouse. He told me I should feel privileged to be in there because "even the women miners aren't allowed in here." I interpreted this incident as another example of masculine-identified territoriality and his identification of me as a provisional visitor.

As one might expect, members of the local community were also curious about what I was doing. One last experience worth recounting here was a conversation I had with an older woman who worked as a maid at the motel where I was staying and getting my meals. One morning she appeared at my door with fresh towels and sheets. For the previous few mornings I had simply exchanged dirty towels for clean ones at the door, declining to let her in the room to clean because it was not necessary and I did not want my scattered books and papers disturbed. On this particular morning, however, she tried to look into the room, at first over my shoulder and then to one side of me. Able to catch only a glimpse of the tangled sheets on the bed, she blurted out, "Honey, you got men in your room at night?" My expression must have reflected my incredulity, because she then broke into gales of laughter. I did, too. But then our conversation became much more serious. After I explained who I was and what I was doing there, she told me, "No man would want you," because someday my paycheck would be too big. She went on to explain that men wanted a woman with a paycheck to help with the bills, but not a paycheck big enough to allow for her independence. Later that week, when she saw me leaving to get in my car, she joked, "You going out to chase after men?" I put my finger to my lips and said, "Shhh! I don't want any competition." She giggled and then disappeared around the corner of the building.

This encounter is significant for several reasons. To begin with, this woman's initial impression of me was based solely on my sex, which to her mind was my most salient feature. The negotiation of my own identity as researcher, as opposed to woman, was accomplished by simply disabusing her of the idea that I was there to exchange sex for money. Indeed, as I discovered later, there was some of that activity occurring at the motel, which partially explained her question. But even more revealing were her comments about the consequences of my "big paycheck" and men's feelings about independent women. In essence, she was saying that men would not only accept but would expect a woman to work outside the home to provide for her family's well-being. However, her paycheck should not rival her husband's. Although men might not be the sole providers in a family, they should remain the principal breadwinners. That way they could continue to occupy a position of greater status, power, and privilege. She was well aware of the expectation that women be economically dependent within the patriarchal family structure according to the prevailing gender ideology.

THE STUDY SITE AND SAMPLE

The coal mine under study was located in southern West Virginia.[4] It began its operations during the slack production period of the mid-1960s, but by the 1990s it was operating twenty-four hours a day to produce more than 3 million tons of coal annually. This mine, like any underground mine, was located in relatively isolated rural area. Mines are classified according to the type of entrance or "portal" used to reach the coal seam as either shaft (vertical tunnel), slope (inclined tunnel), or drift (level horizontal tunnel) mines. In 1990, the mine was a slope mine and miners entered by riding electric-powered steel cars called portabuses. By 1995, its underground mining operations had shifted, and miners entered by descending the shaft by elevator. All miners wore a battery pack that powered the light attached to their hard hats, or "caps," and an oxygen self-rescuer. Like any mine, the inside of this mine was a series of interconnecting and parallel passageways through which miners, their machinery, air, and coal moved in and out. Throughout most of the mine, the ceiling underground, known as the roof or top, was about five and a half feet from the ground. The walls of a mine are called the rib. Miners were working about a mile and a half from the mine entrance.

The procedures for extracting coal underground primarily involve cutting and hauling the coal away from the seam and erecting structures for supporting the roof overhead. In addition, other precautions must be taken to avoid the dangers caused by the removal of coal, such as the leakage of water or methane gas into the mine. As I was told repeatedly by the miners, their mine was particularly "gassy."

4. Bituminous, or soft, coal is the most common kind of coal mined in the United States. It is located in vast seams that lie a few hundred feet beneath the earth's surface and range in thickness from two to more than twelve feet. These seams are concentrated in three regions, known as the Appalachian, Eastern Interior, and Western coal basins. The two eastern coalfields are the principal producers of bituminous coal, which is used for making coke and generating heat and electricity. Since coal was first mined in the United States in the mid-1700s until as recently as 1940, 90 percent or more of all domestically mined coal came from underground mines in the Appalachian region. For several decades after 1940, although underground mining in Appalachia declined in terms of total production, it continued to dominate the other two regions in terms of the total number of miners employed (Perry 1984). In 1987, the Appalachian region contained the bulk of underground mines (1,469) in the United States, followed by the Eastern Interior and Western regions, with sixty and thirty-four mines, respectively (DOE 1987). In the same year, 80 percent of the nation's total underground workforce was located in the Appalachian coalfields.

Work was divided into four shifts in five sections of the mine. Depending on the section, work crews performed either maintenance duties or mined coal using either continuous mining or longwall (plow) methods. Sections and jobs in the mine were thus divided according to either their production or maintenance functions. Each section, whether it was producing coal or was a maintenance or "down" section, was supervised by a section foreman or crew boss, who was a nonunion salaried company employee. Although bosses stayed in close contact with workers, by UMWA contract they were forbidden to operate machinery or otherwise perform any work duties "except in emergencies and except if such work is necessary for the purpose of training or instructing classified employees" (UMWA/BCOA 1988, 5). However, I found out that bosses often pitched in to make the section's production quota. Otherwise, bosses made the day-to-day decisions regarding the production activities, safety, and crew members' work assignments.

As formally outlined in the UMWA/BCOA contracts (1988, 1993), mining jobs were classified in five job grades (see Table 1). Wages, along with the requisite level of mechanical skills and on-the-job responsibilities, increased with the grade of the job. Thus, jobs in Grade 1 generally required fewer skills and more physical strength than jobs in higher grades, which required specific operative skills or certification. General laborers (Grade 1) tended to be the least-skilled workers and occupied the lowest status positions at the mine. At the other end of the scale, a fireboss (Grade 5) was responsible for inspecting each section of the mine for methane gas levels and other safety-related conditions. A fireboss was also responsible for completing the federally required documentation.

At the face, miners used both continuous and longwall mining in a complementary fashion to extract coal. In continuous mining, a single worker operated a thirty-foot-long machine equipped on the front with a clawed rotating drum that tore the coal from the face and loaded it onto a conveyor belt. A longwall machine or "plow" sheared coal from the face while using self-advancing roof jacks to support the roof above it (Perry 1984; Priest 1989). Of all the mining jobs underground, the plow and continuous miner operators had the most responsibility. They had direct contact with a boss, and they were able to act as informal leaders by setting their crew's work pace. The "miner helpers" who did the set-up work for the machine operators also had high standing among crew members. They were not only more skilled workers, per-

Table 1 Coal-Mining Jobs by Grade

Grade		
Grade 5	Continuous Mining Machine Operator	Longwall Machine Operator
	Electrician	Welder, First Class
	Mechanic	Roof Bolter
	Fireboss	
Grade 4	Cutting Machine Operator	Rock Driller
	Dispatcher	Continuous Miner Helper
	Loading Machine Operator	Roof Bolter Helper
	Machine Operator Helper	Maintenance Trainee
	General Inside Repairman and Welder	Electrician Trainee
Grade 3	Driller-Coal	Faceman (including longwall utility man and jacksetter)
	Shooter	Dumper
	Precision Mason-Construction	Shuttle Car Operator
Grade 2	Motorman	Electrician Helper
	Maintenance Trainee	Mechanic Helper
	Electrician Trainee	
Grade 1	Beltman	Pumper
	Bonder	Timberman
	Brakeman	Trackman
	Bratticeman	Wireman
	General Inside Laborer	Laborer-Unskilled
	Mason	

Source: National Bituminous Coal Wage Agreements of 1988/1993

forming duties essential to the cutting of coal, they worked in some of the most dangerous areas in the mine—those with little or no roof support.

Once coal was extracted from the face, it was loaded mechanically into small locomotive cars called "shuttle buggies" and onto long conveyor belts that took the coal out of the mine. The shifting rock overhead always needs to be stabilized to keep the roof from collapsing. Thus, roof bolting, or "pinning top," was crucial to the work process, because it is very unsafe, and also illegal, to work under an unsupported roof. By union contract, a miner can refuse to work in an area she or he deems unsafe. The roof and ribs were supported using either timbers or steel beams. To accomplish this, the roof bolter and her or his assistant would drill and then insert metal rods several feet long, called roof bolts or pins, into the top to bond the thinner layers of strata firmly to the more stable overburden. Meanwhile, buggy or shuttle car operators drove flat motorized cars loaded with coal away from the face and into the main passageway. On their return trip they brought supplies back with them. Other underground workers included electricians and wiremen who hung and maintained communications and cable wire to power the portabuses, mechanics who maintained the machines, and welders who fixed and reinforced metal structures.

Maintenance jobs requiring fewer skills were performed by workers in Grade 1 jobs. Usually working alone or in a group of two to five, these workers maintained mine safety or supported production. For example, section foremen might assign a general inside laborer to hang ventilation curtain, or "rag," pieces of heavy canvas hung in the mine's passageways to let fresh air in and to draw dangerous gasses out. Because production operations raise potentially explosive amounts of coal dust, general laborers were also responsible for "rock dusting," or throwing limestone powder against the ribs to retard coal dust and prevent fires. General inside laborers could also be called upon to lay track and deliver supplies to different sections of the mine. Masons built brick walls for additional roof support and ventilation. The beltmen walked the belt line, shoveling coal that had fallen off the conveyor belts and making sure the belt line was working properly. A pumper checked and adjusted the machines that pump water out of the mine.

At the time of my research, the company employed a dozen assistant foremen or "bosses" and almost five hundred miners. All the bosses were white males. According to the list provided by the company in 1990, there were twenty-three women miners, who constituted almost five percent of the underground labor force. Eighteen of the five hundred miners were black, including three women. By 1996, five of the twenty-three women no longer worked at the mine. Two had been fired and the other three had retired. In 1990, three pairs of women had been assigned to work together permanently. The rest worked solo on otherwise all-male crews. By 1996, the remaining eighteen women miners were working in a solo capacity.

I conducted in-depth interviews with fourteen women miners and had twenty-minute conversations with seven more over a combined period of about two months during the 1990, 1995, and 1996 data collection periods. Seven of the ten interviews done in 1990 were followed up during the summers of 1995 or 1996. Two women I was unable to locate, and a third had died in an accident unrelated to mining. The other four interviews were done during the summers of 1995 or 1996. Most of the women in the sample were hired during the peak period between 1978 and 1980 as a direct result of the industry-wide lawsuit filed by the CEP. Five of these fourteen women were no longer working at the mine at the time of our interviews in either 1995 or 1996. I also did follow-up interviews with the mine superintendent. During each data collection period, I also had either informal conversations or in-depth interviews with seven male miners, three of whom were local union officials at the time.

The sampling of women informants was based on a combination of purposive and snowball techniques. Specifically, I began by asking the first two women I interviewed for the names of and information about other women working at the mine. From the names they provided, I selected other women for interviews on the basis of tenure, job rank, and job-related experiences, such as sexual harassment or alleged discrimination. When setting up these interviews, I either called them on the telephone or met them in the women's bathhouse, usually before their shift. Because the women and men miners I interviewed provided a variety of perspectives on topics relevant to socialist feminist theory, I continued to interview them until I reached a point of redundancy in my data collection (Lincoln and Guba 1985).

The women I interviewed in 1990, 1995, and 1996 were diverse in age, education, marital status, childbearing, mining experience, and current job. When they were hired, between 1978 and 1980, most of the fourteen women were in their twenties or early thirties. By 1996, the majority were in their late thirties and forties. Only two women were older. One was fifty-six and the other, now retired, was sixty-two. Two had finished sixth grade, another had finished tenth grade, nine had high school diplomas, and two had attended college but never graduated. Seven were either single or divorced with children when they began mining. Four more divorced after they were hired and also had children to support. Two were single with no children. Only one was married with children when she was hired and stayed married up to the time of the 1995 interview. All of them said they need a coal-mining job to support either themselves or their families. Two of the women were black and the rest were white. The women's mining experience ranged from twelve to twenty-one years. At the time of my last interview with them, nine of the fourteen women were working in Grade 1 jobs as general inside laborers, beltmen, masons, or wiremen. The other five had held jobs in one of the higher grades, including motorman (Grade 2), jacksetter, long-wall utility man, and shuttle car operator (Grade 3), and roof bolter (Grade 5). It should be noted that of the nine women in Grade 1 jobs, three had their firebossing papers in 1990 and three more had the same certification by 1995. At any time any one of them could be, and often was, temporarily reassigned to that Grade 5 position. In 1990, two of the fourteen women had been working together, but by 1995, as mentioned previously, all of them were working as the sole woman on their crews.

Unless otherwise indicated, interviews with miners, especially the women, lasted at least three hours. Most of the interviews took place in my motel room or in miners' homes, while others took place in the women's bathhouse, at a local diner, or in isolated places on the company's property. These more private locations were selected so that informants felt less inhibited about discussing sensitive topics. All but two interviews were tape-recorded; otherwise extensive notes were taken. One woman was interviewed in a diner, so I elected not to tape the interview and call attention to us. The mine superintendent did not grant me permission to tape either of my two interviews with him. During all of my interviews with women, I made a point of bringing several props, which were often used for reference or might elicit further conversation and provide me a greater understanding of what they were telling me. These items were a copy of the current union contract, a list the company sent me of the women it employed, and some materials with information about the CEP. Observation was done during my frequent visits to the women's bathhouse and occasional visits to the mine office, during my underground tour of the mine, and out in the community. I took field notes daily, particularly after interviews or contact with other members of the mining community. Document study was limited to recording data from open records posted in the lamphouse and included data about all miners' start dates, lengths of service, and current job status.

My analysis of the data was done both during and after my fieldwork. Tapes were transcribed either in the field or sometime thereafter, yielding several hundred pages of information. Both anticipated and unanticipated themes emerged from the transcribed interviews, field notes, and documents. Some coding categories came from the theoretical framework (i.e., experiences with sexual harassment or on-the-job training). Others emerged from the recurrence of reported events and perceptions. These inductively derived categories were then compared with the predicted theoretical ones, which informed the organization of the research results. This is the logic of the "pattern-matching" mode of qualitative analysis discussed by numerous qualitative methodologists (Marshall and Rossman 1995; Miles and Huberman 1994; Stake 1995; Yin 1987). The "constant comparison" of these data (Glaser and Strauss 1967) yielded a general picture used to assess the value of the theoretical framework and corresponding concepts.

REFERENCES

Althouse, Ron. 1974. "Work, Safety, and Lifestyle Among Southern Appalachian Coal Miners: A Survey of the Men of Standard Mines." *West Virginia University Bulletin Series* 74, no. 11-9. Morgantown: West Virginia University Press.

Anglin, Mary. 1983. "Experiences of In-migrants in Appalachia." In *Appalachia and America: Autonomy and Regional Dependence*, ed. Allan Batteau, 227–38. Lexington: University Press of Kentucky.

———. 1993. "Engendering the Struggle: Women's Labor and Traditions of Resistance in Rural Southern Appalachia." In *Fighting Back in Appalachia: Traditions of Resistance and Change*, ed. Stephen L. Fisher, 263–82. Philadelphia: Temple University Press.

———. 2002. *Women, Power, and Dissent in the Hills of Carolina*. Urbana: University of Illinois Press.

Appalachian Alliance. 1982. *Appalachia in the Eighties: A Time for Action*. New Market, Tenn.: Appalachian Alliance.

Aptheker, Bettina. 1989. *Tapestries of Life: Women's Consciousness and the Meaning of Daily Experience*. Amherst: University of Massachusetts Press.

Ayers, Edward L. 1992. *The Promise of the New South: Life After Reconstruction*. New York: Oxford University Press.

Banks, Alan, Dwight Billings, and Karen Tice. 1993. "Appalachian Studies, Resistance, and Postmodernism." In *Fighting Back in Appalachia: Traditions of Resistance and Change*, ed. Stephen L. Fisher, 283–302. Philadelphia: Temple University Press.

Barry, Joyce. 2001. "Mountaineers Are Always Free? An Examination of the Effects of Mountaintop Removal in West Virginia." *Women's Studies Quarterly* 29 (1–2): 116–30.

Beaver, Patricia D. 1999. "Women in Appalachia and the South: Gender, Race, Region, and Agency." *National Women's Studies Association Journal* 11 (3): ix–xxix.

Billings, Dwight B., and Kathleen M. Blee. 2000. *The Road to Poverty: The Making of Wealth and Hardship in Appalachia*. Cambridge: Cambridge University Press.

Billings, Dwight B., Kathleen M. Blee, and Louis E. Swanson. 1986. "Culture, Family, and Community in Preindustrial Appalachia." *Appalachian Journal* 13 (2): 154–70.

Billings, Dwight B., and Ann R. Tickamyer. 1993. "Uneven Development in Appalachia." In *Forgotten Places: Uneven Development in Rural America,* ed. Thomas A. Lyson and William W. Falk, 7–29. Lawrence: University Press of Kansas.

Blauner, Robert. 1969. "Internal Colonialism and Ghetto Revolt." *Social Problems* 16 (4): 393–408.

Bokemeier, Janet L., and Ann R. Tickamyer. 1985. "Labor Force Experiences of Nonmetropolitan Women." *Rural Sociology* 50 (1): 51–73.

Brueggebors, Barbara E. 1979. "Miner Dives to Safety, But . . . Tragic Scene Etched in His Mind." State College, Pennsylvania, *Centre Daily Times,* October 6.

Business Week. 1979. "The Militant Women Mining the Coalfields." June 25, p. 30.

Butani, Shail J., and Ann M. Bartholomew. 1988. "Characterization of the 1986 Coal Mining Workforce." Bureau of Mines Information Circular 9192. Washington, D.C.: U.S. Department of the Interior, Bureau of Mines.

Cable, Sherry. 1993. "From Fussin' to Organizing: Individual and Collective Resistance at Yellow Creek." In *Fighting Back in Appalachia: Traditions of Resistance and Change,* ed. Stephen L. Fisher, 69–84. Philadelphia: Temple University Press.

Caudill, Harry M. 1962. *Night Comes to the Cumberlands: A Biography of a Depressed Area.* Boston: Little, Brown.

Coal Age. 1986. "Job Discrimination: Women Miners Dig Away at Discrimination." August, pp. 11–13.

Cockburn, Cynthia. 1983. *Brothers: Male Dominance and Technological Change.* London: Pluto Press.

———. 1985. *Machinery of Dominance: Women, Men, and Technical Know-how.* Dover, N.H.: Pluto Press.

———. 1991. *In the Way of Women: Men's Resistance to Sex Equality in Organizations.* Ithaca, N.Y.: ILR Press.

Collins, Patricia Hill. 1990. *Black Feminist Thought: Knowledge, Consciousness, and the Politics of Empowerment.* New York: Routledge.

Couto, Richard. 1993. "The Memory of Miners and the Conscience of Capital: Coal Miners' Strikes as Free Spaces." In *Fighting Back in Appalachia: Traditions of Resistance and Change,* ed. Stephen L. Fisher, 165–94. Philadelphia: Temple University Press.

Dawson, Kipp. 1990. "Your Sisters Underground: Part 1." *Sojourner.* May, pp. 20–21.

———. 1992. "Women Miners and the UMWA: 1973–1983." Occasional Paper no. 11. New York: City University of New York, Center for Labor-Management Policy Studies.

Deaux, Kay. 1983. *Women of Steel.* New York: Praeger.

Denzin, Norman K. 1989. *The Research Act: A Theoretical Introduction to Sociological Methods.* Englewood Cliffs, N.J.: Prentice-Hall.

Dix, Keith. 1977. *Work Relations in the Coal Industry: The Handloading Era, 1880–1930.* Morgantown: West Virginia University Press.

———. 1988. *What's a Coal Miner to Do? The Mechanization of Coal Mining.* Pittsburgh: University of Pittsburgh Press.

Duncan, Cynthia M. 1999. *Worlds Apart: Why Poverty Exists in Rural America*. New Haven: Yale University Press.

Eisenstein, Zillah H. 1984. *Feminism and Sexual Equality: Crisis in Liberal America*. New York: Monthly Review Press.

Eller, Ronald D. 1982. *Miners, Millhands, and Mountaineers: Industrialization of the Appalachian South, 1880–1930*. Knoxville: University of Tennessee Press.

Enarson, Elaine Pitt. 1984. *Woods-Working Women: Sexual Integration in the U.S. Forest Service*. Birmingham: University of Alabama Press.

England, Lynn, and Ralph B. Brown. 2003. "Community and Resource Extraction in Rural America." In *Challenges for Rural America in the Twenty-first Century*, ed. David L. Brown and Louis E. Swanson, 317–28. University Park: Pennsylvania State University Press.

Evans, Sara M., and Harry C. Boyte. 1986. *Free Spaces: The Sources of Democratic Change in America*. New York: Harper and Row.

Eviota, Elizabeth. 1992. *The Political Economy of Gender*. London: Zed Books.

Felski, Rita. 1989. "Feminist Theory and Social Change." *Theory, Culture, and Society* 6: 219–40.

Fielding, Nigel G., and Jane L. Fielding. 1986. *Linking Data*. Qualitative Research Methods Series no. 4. Beverly Hills, Calif.: Sage Publications.

Fleischauer, Barbara Evans. 1989a. "Update on Salyers West Virginia Discrimination Case." *Coal Mining Women's Support Team News* 11 (4): 11.

———. 1989b. "Women Miner Denied Damages, But West Virginia Supreme Court Agrees to Reconsider Salyers Case." *Coal Mining Women's Support Team News* 11 (2): 1–12.

Flora, Cornelia B., and Sue Johnson. 1978. "Discarding the Distaff: New Roles for Rural Women." In *Rural USA: Persistence and Change*, ed. Thomas Ford. Ames: Iowa State University Press.

Fonow, Mary Margaret. 2003. *Union Women: Forging Feminism in the United Steelworkers of America*. Minneapolis: University of Minnesota Press.

Freire, Paulo. 1972. *Pedagogy of the Oppressed*. New York: Herder and Herder.

Gaventa, John. 1980. *Power and Powerlessness: Quiescence and Rebellion in an Appalachian Valley*. Urbana: University of Illinois Press.

Gibson-Graham, J. K. 1996. *The End of Capitalism (as We Knew It): A Feminist Critique of the Political Economy*. Cambridge: Blackwell.

Giesen, Carol A. B. 1995. *Coal Miners' Wives: Portraits of Endurance*. Lexington: University Press of Kentucky.

Glaser, B. G., and A. L. Strauss. 1967. *The Discovery of Grounded Theory: Strategies for Qualitative Research*. Chicago: Aldine.

Goffman, Erving. 1959. *The Presentation of Self in Everyday Life*. Garden City, N.Y.: Doubleday.

Green, James R. 1996. "'Tying the Knot of Solidarity': The Pittston Strike of 1989–1990." In *The United Mine Workers of America: A Model of Industrial Solidarity?* ed. John H. M. Laslett, 513–44. University Park: Pennsylvania State University Press.

Gruber, James S., and Lars Bjorn. 1982. "Blue-Collar Blues: The Sexual Harassment of Women Autoworkers." *Work and Occupations* 9 (3): 271–98.

Gutek, Barbara A. 1985. *Sex and the Workplace: The Impact of Sexual Behavior and Harassment on Women, Men, and Organizations.* San Francisco: Jossey Bass.

Hall, Betty Jean. 1984. "Coal Mining Women Confront Pattern of Discrimination." *Mountain Life and Work* 60 (7): 4–9.

———. 1990. "Women Miners Can Dig It, Too!" In *Communities in Economic Crisis: Appalachia and the South,* ed. John Gaventa, Barbara Ellen Smith, and Alex Willingham, 53–60. Philadelphia: Temple University Press.

Hammond, Judith, and Constance W. Mahoney. 1983. "Reward-Cost Balancing Among Women Coal Miners." *Sex Roles* 9 (1): 17–29.

Haney, Wava G. 1982. "Women." In *Rural Society in the U.S.: Issues for the 1980s,* ed. Don A. Dillman and Darly J. Hobbs, 124–35. Boulder: Westview Press.

Haraway, Donna J. 1991. *Simians, Cyborgs, and Women: The Reinvention of Nature.* New York: Routledge.

Harding, Sandra. 1986. *The Science Question in Feminism.* Ithaca: Cornell University Press.

———. 1987. *Feminism and Methodology: Social Science Issues.* Bloomington: Indiana University Press.

Harris, Rosalind P., Jeffery C. Bridger, Carolyn Sachs, and Suzanne E. Tallichet. 1995. "Empowering Rural Sociology: Exploring and Linking Alternative Paradigms in Theory and Methodology." *Rural Sociology* 60 (4): 585–606.

Hartmann, Heidi. 1976. "Capitalism, Patriarchy, and Job Segregation by Sex." In *Women and the Workplace: The Implications of Occupational Segregation,* ed. Martha Blaxall and Barbara Reagan, 137–69. Chicago: University of Chicago Press.

———. 1984. "The Unhappy Marriage of Marxism and Feminism: Towards a More Progressive Union." In *Feminist Frameworks: Alternative Theoretical Accounts of the Relations Between Women and Men,* 2d ed., ed. Allison M. Jaggar and Paula S. Rothenberg, 172–211. New York: McGraw-Hill.

Hartsock, Nancy. 1983. *Money, Sex, and Power: Toward a Feminist Historical Materialism.* New York: Longman.

Howes, Candace, and Ann R. Markusen. 1981. "Poverty: A Regional Political Economy Perspective." In *Nonmetropolitan America in Transition,* ed. Amos Hawley and Sara Mills Maize, 437–63. Chapel Hill: University of North Carolina Press.

Jaggar, Allison M. 1988. *Feminist Politics and Human Nature.* Totowa, N.J.: Rowman and Allanheld.

Kanter, Rosabeth Moss. 1976. "The Impact of Hierarchical Structures on the Work Behavior of Men and Women." *Social Problems* 23 (4): 415–30.

———. 1977. *Men and Women of the Corporation.* New York: Harper and Row.

Keerdoja, Eileen, Kim Foltz, and Joyce Barnathan. 1982. "Women in the Coal Mines: The Struggle Goes On." *Newsweek,* April 19, pp. 18–19.

Kingsolver, Ann E. 1992. "Five Women Negotiating the Meaning of Negotiation." *Anthropology Quarterly* 65: 101–4.

Laslett, John H. M. 1996. *The United Mine Workers of America: A Model of Industrial Solidarity?* University Park: Pennsylvania State University Press.

Lewis, Helen M. 1967. "The Subcultures of the Southern Appalachians, Their Origins and Boundary Maintenance." Paper presented at the Institute of Southern Culture, Longwood College, Farmville, Virginia, February 17.

———. 1970. "Coal Miners' Peer Groups and Family Roles." Paper presented at the American Anthropological Association Annual Meeting, San Diego, California, November 22.

Lewis, Helen M., Linda Johnson, and Don Askins, eds. 1978. *Colonialism in Modern America: The Appalachian Case.* Boone, N.C.: Appalachian Consortium Press.

Lewis, Helen M., and Edward E. Knipe. 1978. "The Colonialism Model: The Appalachian Case." In *Colonialism in Modern America: The Appalachian Case,* ed. Helen M. Lewis, Linda Johnson, and Don Askins, 9–31. Boone, N.C.: Appalachian Consortium Press.

Lewis, Ronald. 1987. *Black Coal Miners in America: Race, Class, and Community Conflict, 1780–1980.* Lexington: University Press of Kentucky.

———. 1989. "From Peasant to Proletarian: The Migration of Southern Blacks to the Central Appalachian Coalfields." *Journal of Southern History* 60 (1): 77–102.

Lilly, Leslie. 1989. "Training Women for Jobs in Rural Economics: A Southern Experience." In

Job Training for Women: The Promise and Limits of Public Policies, ed. Sharon L. Harlan and Ronnie J. Steinberg, 247–68. Philadelphia: Temple University Press.

Lincoln, Yvonne S., and Egon G. Guba. 1985. *Naturalistic Inquiry.* Beverly Hills, Calif.: Sage Publications.

Lipton, Michael. 2002. "The Fight for the Soul of Coal Country." *New York Times,* May 17, A25.

Lorde, Audre. 1984. *Sister Outside.* Trumansberg, N.Y.: Crossing Press.

Lukes, Stephen. 1974. *Power: A Radical View.* London: Macmillan.

MacKinnon, Catherine. 1979. *Sexual Harassment of Working Women.* New Haven: Yale University Press.

Maggard, Sally Ward. 1981. "From Farmers to Miners: The Decline of Agriculture in Eastern Kentucky." In *Science and Agricultural Development,* ed. Lawrence Busch, 25–66. Totowa, N.J.: Allanheld, Osum and Co.

———. 1985. "Class and Gender: New Theoretical Priorities in Appalachian Studies." In *The Impact of Institutions in Appalachia,* ed. Jim Lloyd and Anne G. Campbell, 100–113. Boone, N. C.: Appalachian Consortium Press.

———. 1990 "Gender Contested: Women's Participation in the Brookside Coal Strike." In *Women and Social Protest,* ed. Guida West and Rhoda Lois Blumberg, 75–90. New York: Oxford University Press.

———. 1994. "From Farm to Coal Camp to Back Office and McDonald's: Living in the Midst of Appalachia's Latest Transformation." *Journal of the Appalachian Studies Association* 6 (1): 14–28.

———. 1998. "We're Fighting Millionaires! The Clash of Gender and Class in Appalachian Women's Union." In *No Middle Ground: Women and Radical Protest,* ed. Kathleen M. Blee, 289–306. New York: New York University Press.

———. 1999. "Gender, Race, and Place: Confounding Labor Activism in Central

Appalachia." In *Neither Separate nor Equal: Women, Race, and Class in the South*, ed. Barbara Ellen Smith, 185–206. Philadelphia: Temple University Press.

Mahoney, Constance. 1978. Appalachian Women's Perception of Their Work Experiences as Underground Coal Miners. Master's thesis, East Tennessee State University.

Marshall, Catherine, and Gretchen B. Rossman. 1995. *Designing Qualitative Research*. 2d ed. Thousand Oaks, Calif.: Sage Publications.

Martin, Susan. 1980. *Breaking and Entering: Policewomen on Patrol*. Berkeley and Los Angeles: University of California Press.

McGranahan, David A. 2003. "How People Make a Living in Rural America." In *Challenges for Rural America in the Twenty-first Century*, ed. David L. Brown and Louis E. Swanson, 135–51. University Park: Pennsylvania State University Press.

McIlwee, Judith. 1981. "Organization Theory and the Entry of Women into Nontraditional Occupations." *Western Sociological Review* 12 (1): 33–52.

Meyer, Herbert H., and Mary Dean Lee. 1978. "Women in Traditionally Male Occupations: The Experience of Ten Public Utilities Companies." Washington, D.C.: U.S. Government Printing Office.

Mies, Maria. 1986. *Patriarchy and Accumulation on a World Scale: Women in the International Division of Labor*. London: Zed Press.

Miles, Matthew B., and A. Michael Huberman. 1994. *Qualitative Data Analysis: An Expanded Sourcebook*. 2d ed. Thousand Oaks, Calif.: Sage Publications.

Milkman, Ruth. 1985. "Women Workers, Feminism and the Labor Movement Since the 1960s." In *Women, Work, and Protest: A Century of U.S. Women's Labor History*, ed. Ruth Milkman, 300–322. Boston: Routledge and Kegan Paul.

———. 1987. *Gender at Work: The Dynamics of Job Segregation by Sex During World War II*. Urbana: University of Illinois Press.

Mills, C. Wright. 1959. *The Sociological Imagination*. New York: Oxford University Press.

Moore, Marat. 1996a. "Women Go Underground." In *The United Mine Workers of America: A Model of Industrial Solidarity?* ed. John H. M. Laslett, 484–512. University Park: Pennsylvania State University Press.

———. 1996b. *Women in the Mines: Stories of Life and Work*. New York: Twayne Publishers.

Moser, Caroline. 1989. "Gender Planning in the Third World: Meeting Practical and Strategic Gender Needs." *World Development* 17 (11): 1799–1825.

Mountain Life and Work. 1978a. "Women Push Coal Employment." August, pp. 22–23.

———. 1978b "Women Sue 153 Coal Companies." June, pp. 3–6.

———. 1979 "Women Rally at National Conference." July/August, pp. 4–9.

———. 1984 "Women in Nontraditional Jobs in Appalachia." September (special issue).

Naples, Nancy. 1998. "Women's Community Activism and Feminist Activist Research." In *Community Activism and Feminist Politics: Organizing Across Race, Class, and Gender*, ed. Nancy Naples, 1–27. New York: Routledge.

New York Times. 1982. "Women Miners Seek Damages in Harassment Suit." April 29, A24.

Nielsen, Joyce McCarl. 1990. "Introduction." In *Feminist Research Methods: Exemplary Readings in the Social Sciences,* ed. Joyce McCarl Neilsen, 1–37. Boulder: Westview Press.

Oberhauser, Ann M., and Anne-Marie Turnage. 1999. "A Coalfield Tapestry: Weaving the Socio-economic Fabric of Women's Lives." In *Neither Separate nor Equal: Women, Race, and Class in the South,* ed. Barbara Ellen Smith, 185–206. Philadelphia: Temple University Press.

O'Farrell, Brigid. 1982. "Women and Nontraditional Blue-collar Jobs in the 1980s: An Overview." In *Women in the Workplace,* ed. Phyllis A. Wallace, 135–65. Boston: Auburn House.

———. 1988. "Women in Blue-collar Occupations: Traditional and Nontraditional." In *Women Working: Theories and Facts in Perspective,* ed. Ann Helton Stromberg and Shirley Harkess, 258–72. Mountain View, Calif.: Mayfield Publishing.

O'Farrell, Brigid, and Sharon Harlan. 1982. "Craftsworkers and Clerks: The Effect of Male Co-Worker Hostility on Women's Satisfaction with Nontraditional Blue-Collar Jobs." *Social Problems* 26 (3): 252–64.

Ollenberger, Jane C., and Helen A. Moore. 1998. *A Sociology of Women: The Intersection of Patriarchy, Capitalism, and Colonization.* 2d ed. Upper Saddle River, N.J.: Prentice-Hall.

Ong, Aihwa. 1987. *Spirits of Resistance and Capitalist Discipline: Factory Women in Malaysia.* Albany: State University of New York Press.

Padavic, Irene. 1989. "Attractions of Male Blue-Collar Jobs for Black and White Women: Economic Need, Exposure, and Attitudes." *Social Science Quarterly* 72: 33–49.

Padavic, Irene, and Barbara F. Reskin. 1990. "Men's Behavior and Women's Interest in Blue-Collar Jobs." *Social Problems* 37 (4): 613–28.

Parker, H. Jane. 1992. "Engendering Identity(s) in a Rural Arkansas Ozark Community." *Anthropology Quarterly* 65: 148–55.

Perry, Charles R. 1984. *Collective Bargaining and the Decline of the United Mine Workers of America.* Industrial Research Unit no. 60. Philadelphia: University of Pennsylvania Press.

President's Commission on Coal. 1980. *The American Coal Miner: A Report on Community and Living Conditions in the Coal Fields.* Washington, D.C.: U.S. Government Printing Office.

Priest, Dana. 1989. "Below Ground in Coal Country." *Washington Post,* November 26, Metro section.

Pudup, Mary Beth. 1990. "Women's Work in the West Virginia Economy." *West Virginia History* 49: 7–20.

Punch, Maurice. 1986. *The Politics and Ethics of Fieldwork.* Qualitative Research Methods Series no. 3. Beverly Hills, Calif.: Sage Publications.

Reskin, Barbara F. 1988. "Bringing the Men Back In: Sex Differentiation and the Devaluation of Women's Work." *Gender and Society* 2 (1): 58–81.

———. 1993. "Sex Segregation in the Workplace." In *Annual Review of Sociology,* vol. 19, ed. Judith Blake and John Hagen, 241–70. Palo Alto, Calif.: Annual Reviews, Inc.

Reskin, Barbara F., and Heidi Hartmann. 1986. *Women's Work, Men's Work: Sex Segregation on the Job*. Washington, D.C.: National Academy Press.

Reskin, Barbara F., and Irene Padavic. 1988. "Supervisors as Gatekeepers: Male Supervisors' Response to Women's Integration in Plant Jobs." *Social Problems* 35 (5): 536–50.

Reskin, Barbara F., and Patricia A. Roos. 1987. "Sex Segregation and Status Hierarchies." In *Ingredients for Women's Employment Policy*, ed. Christine Bose and Glenna Spitze, 1–21. Albany: State University of New York Press.

Rogers, William P. 1978. *The Coal Primer: A Handbook for the Non-coal Person*. Van Buren, Ark.: Valley Press.

Ross, M. H. 1974. "Lifestyle of the Coal Miner: America's First Hard Hat." In *Humanizing the Workplace*, ed. Roy P. Fairfield, 171–80. New York: Prometheus Books.

Sachs, Carolyn. 1996. *Gendered Fields: Rural Women, Agriculture, and Environment*. Boulder: Westview Press.

Salstrom, Paul. 1994. *Appalachia's Path to Dependency: Rethinking a Region's Economic History, 1730–1940*. Lexington: University Press of Kentucky.

Schur, Edwin M. 1984. *Labeling Women Deviant: Gender, Stigma, and Control*. Philadelphia: Temple University Press.

Scott, James C. 1985. *Weapons of the Weak: Everyday Forms of Peasant Resistance*. New Haven: Yale University Press.

Scott, Shaunna L. 1995. *Two Sides to Everything: The Cultural Construction of Class Consciousness in Harlan County, Kentucky*. Albany: State University of New York Press.

Seitz, Virginia Rinaldo. 1995. *Women, Development, and Communities for Empowerment in Appalachia*. Albany: State University of New York Press.

Shapiro, Henry. 1978. *Appalachia on Our Minds: The Southern Mountains and Mountaineers in American Consciousness, 1970–1920*. Chapel Hill: University of North Carolina Press.

Sheehan, Charles. 2004. "Coal Companies Digging Deep to Lure Miners." *Centre Daily Times*, October 12.

Shifflett, Crandall A. 1991. *Coal Towns: Life, Work, and Culture in Company Towns of Southern Appalachia, 1880–1960*. Knoxville: University of Tennessee Press.

Silverman, Dierdre. 1976. "Sexual Harassment: Working Women's Dilemma." *Quest: A Feminist Quarterly* 3: 15–24.

Simon, Richard M. 1983. "Hard Times for Organized Labor in Appalachia." *Review of Radical Political Economics* 15 (3): 21–34.

Smith, Barbara Ellen. 1999. "'Beyond the Mountains': The Paradox of Women's Place in Appalachian History." *National Women's Studies Association Journal* 11 (3): 1–18.

Smith, Dorothy. 1987. *The Everyday World as Problematic: A Feminist Sociology*. Toronto: University of Toronto Press.

———. 1990. *Conceptual Practices of Power*. Boston: Northeastern University Press.

Sokoloff, Natalie J. 1988. "Contributions of Marxism and Feminism to the Sociol-

ogy of Women and Work." In *Women Working: Theories and Facts in Perspective,* ed.
Ann Helton Stromberg and Shirley Harkess, 116–31. Mountain View, Calif.: Mayfield Publishing.

Stake, Robert E. 1995. *The Art of Case Study Research.* Thousand Oaks, Calif.: Sage Publications.

Strober, Myra. 1984. "Toward a General Theory of Occupational Sex Segregation: The Case of Public School Teaching. In *Sex Segregation in the Workplace: Trends, Explanations, and Remedies,* ed. Barbara F. Reskin, 144–56. Washington, D.C.: National Academy Press.

Swerdlow, Marian. 1989. "Entering a Nontraditional Occupation: A Case of Rapid Transit Operatives." *Gender and Society* 3 (3): 373–87.

Tallichet, Suzanne E. 1991. "Moving Up Down in the Mine: Sex Segregation Among Underground Coal Miners." Ph.D. dissertation, Pennsylvania State University, Department of Agricultural Economics and Rural Sociology.

———. 1995. "Gendered Relations in the Mines and the Division of Labor Underground." *Gender and Society* 9 (6): 697–711.

Thompson, Alexander M., III. 1979. *Technology, Labor, and Industrial Structure of the U.S. Coal Industry: A Historical Perspective.* New York: Garland.

Thrasher, Sue. 1981. "Coal Employment Project." *Southern Exposure* 4 (winter): 47–51.

Tickamyer, Ann R. 1996. "Sex, Lies, and Statistics: Can Rural Sociology Survive Restructuring? (Or) What Is Right with Rural Sociology and How Can We Fix It?" *Rural Sociology* 61 (1): 5–24.

Tickamyer, Ann R., and Debra A. Henderson. 2003. "Rural Women: New Roles for the New Century?" In *Challenges for Rural America in the Twenty-first Century,* ed. David L. Brown and Louis E. Swanson, 109–17. University Park: Pennsylvania State University Press.

Tomaskovic-Devey, Donald. 1993. *Gender and Racial Inequality at Work.* Ithaca, N.Y.: ILR Press.

Trent, Roger B., and Nancy Stout-Wiegand. 1987. "Attitudes Toward Women Coal Miners in an Appalachian Coal Community." *Journal of Community Development Society* 18 (1): 1–14.

Trotter, J. W., Jr. 1990. *Coal, Class, and Color: Blacks in Southern West Virginia, 1915–32.* Urbana: University of Illinois Press.

Turner, William H., and Edward J. Cabbell. 1985. *Blacks in Appalachia.* Lexington: University Press of Kentucky.

Uggen, Christopher, and Amy Blackstone. 2004. "Sexual Harassment as a Gendered Expression of Power." *American Sociological Review* 69 (1): 64–92.

United Mine Workers' Journal. 1979a. "Today's UMWA Miners: Banking on a New Image." January/February, pp. 22–25.

———. 1979b. "Union Sponsors Women's Conference, First in History of UMWA." November, pp. 8–9.

———. 1980a. "Women Miners Increasing." July, p. 12.

———. 1980b. "Women Must Learn to Cope with Unique Problems." July, pp. 10–12.

———. 1982a. "Health, Safety Tops Women's Miners' List." February 16–28, p. 7.
———. 1982b. "Ida Mae Stull, Intent, Fought for Her Rights." February 1–15, pp. 12–13.
———. 1982c. "Women Miners Urge Solidarity." August 1–15, p. 7.
———. 1983. "What We Want Is Equal Treatment." July/August, pp. 14–16.
———. 1984. "Women Miners' Conference: We're Working to Strengthen the Union." July, pp. 16–17.
United Mine Workers of America/Bituminous Coal Operators of America (UMWA/BCOA). 1988. *National Bituminous Coal Wage Agreement.* Indianapolis: Allied Printing.
———. 1993. *National Bituminous Coal Wage Agreement.* Indianapolis: Allied Printing.
US. Department of Energy (DOE). 1980. *Coal Production Annual.* DOE/EIA-0118 (80). Washington, D.C.: U.S. Government Printing Office.
———. 1987. *Coal Production Annual.* DOE/EIA-0118 (87). Washington, D.C.: U.S. Government Printing Office.
US. Equal Employment Opportunity Commission (EEOC). 1979. *Employment Analysis Report. EEO-1 Report Summary of Nationwide Industries (Bituminous Coal and Lignite Mining).* Washington, D.C.: U.S. Government Printing Office.
Vaught, Charles, and David L. Smith. 1980. "Incorporation and Mechanical Solidarity in an Underground Mine." *Sociology of Work and Occupations* 7 (2): 159–87.
Vidich, Arthur J., and Stanley M. Lyman. 1994. "Qualitative Methods: Their History in Sociology and Anthropology." In *Handbook of Qualitative Research,* ed. Norman K. Denzin and Yvonne S. Lincoln, 41–110. Thousand Oaks, Calif.: Sage Publications.
Walls, David S., and Dwight B. Billings. 1977. "The Sociology of Southern Appalachia." *Appalachian Journal* 5 (1): 131–44.
Walshok, Mary L. 1981. *Blue Collar Women: Pioneers on the Male Frontier.* New York: Anchor Books.
Wardwell, Mark L., Charles Vaught, and David L. Smith. 1985. "Underground Coal Mining and the Labor Process." In *The Rural Workforce: Non-agricultural Occupations in America,* ed. Clifford D. Bryant, Donald J. Shoemaker, James K. Skipper Jr., and William E. Snizek, 43–61. South Hadley, Mass.: Bergin and Garvey.
Warren, Carol A. B. 1988. *Gender Issues in Field Research.* Qualitative Research Methods Series no. 9. Beverly Hills, Calif.: Sage Publications.
Weatherspoon, Floyd D. 1985. *Equal Employment Opportunity and Affirmative Action: A Sourcebook.* New York: Garland.
Weiss, Chris. 1993. "Appalachian Women Fight Back: Organizational Approaches to Nontraditional Job Advocacy." In *Fighting Back in Appalachia: Traditions of Resistance and Change,* ed. Stephen L. Fisher, 151–64. Philadelphia: Temple University Press.
Weller, Jack E. 1965. *Yesterday's People: Life in Contemporary Appalachia.* Lexington: University Press of Kentucky.

West, Candace, and Don H. Zimmerman. 1987. "Doing Gender." *Gender and Society* 1 (2): 125–251.

Westkott, Marcia. 1990. "Feminist Criticism of the Social Sciences." In *Feminist Research Methods: Exemplary Readings in the Social Sciences,* ed. Joyce McCarl Neilsen, 58–68. Boulder: Westview Press.

Whatmore, Sarah, Terry Mardsen, and Phillip Lowe. 1994. *Gender and Rurality.* London: David Fulton.

Whisnant, David E. 1980. *Modernizing the Mountaineer: People, Power, and Planning in Appalachia.* New York: Burt Franklin and Co.

White, Connie, Barbara Angle, and Marat Moore. 1981. *Sexual Harassment in the Coal Industry: A Survey of Women Miners.* Oak Ridge, Tenn.: Coal Employment Project.

White, Steven E. 1998. "Migration Trends in the Kansas Ogallala Region and the Internal Colonial Dependency Model." *Rural Sociology* 63 (2): 253–71.

Wilayto, Phil, and Dave Cormier. 1990. *'We Won't Go Back': The Story of Struggle of the United Mine Workers of America Against the Pittston Coal Company.* New York: United Labor Action.

Woods, Ellen Meiksins. 1995. *Democracy Against Capitalism: Renewing Historical Materialism.* Cambridge: Cambridge University Press.

Yarrow, Michael. 1985. "Capitalism, Patriarchy, and 'Men's Work': The System of Control of Production of Coal Mining." In *The Impact of Institutions in Appalachia,* ed. Jim Lloyd and Anne G. Campbell, 29–47. Boone, N.C.: Appalachian Consortium Press.

———. 1990. "Voices from the Coalfields: How Miners' Families Understand the Crisis of Coal." In *Communities in Economic Crisis: Appalachia and the South,* ed. John Gaventa, Barbara Ellen Smith, and Alex Willingham, 38–52. Philadelphia: Temple University Press.

———. 1991. "The Gender-Specific Class Consciousness of Appalachian Coal Miners: Structure and Change." In *Bringing Class Back In: Contemporary and Historical Perspectives,* ed. Scott G. McNall, Rhonda F. Levine, and Rick Fantasia, 285–310. Boulder: Westview Press.

Yin, Robert K. 1987. *Case Study Research: Design and Methods.* Beverly Hills, Calif.: Sage Publications.

Yount, Kristen. 1986. "Women and Men Coal Miners: Coping with Gender Integration Underground." Ph.D. dissertation, University of Colorado, Boulder, Department of Sociology.

INDEX

RURAL STUDIES SERIES

Clare Hinrichs, General Editor

The Estuary's Gift:
An Atlantic Coast Cultural Biography
David Griffith

Sociology in Government:
The Galpin-Taylor Years in the
U.S. Department of Agriculture, 1919–1953
Olaf F. Larson and Julie N. Zimmerman
Assisted by Edward O. Moe

Challenges for Rural America in the Twenty-First Century
Edited by David L. Brown and Louis Swanson

A Taste of the Country:
A Collection of Calvin Beale's Writings
Peter A. Morrison

Farming for Us All:
Practical Agriculture and the Cultivation of Sustainability
Michael Mayerfeld Bell

Together at the Table:
Sustainability and Sustenance in the
American Agrifood System
Patricia Allen

Country Boys:
Masculinity and Rural Life
Edited by Hugh Campbell,
Michael Mayerfeld Bell, and Margaret Finney

Welfare Reform in Persistent Rural Poverty:
Dreams, Disenchantments, and Diversity
Kathleen Pickering, Mark H. Harvey,
Gene F. Summers, and David Mushinski